Lukas König

Complex Behavior in Evolutionary Robotics

Also of interest

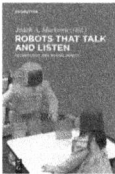

Robots that Talk and Listen
Judith Markowitz (Ed.), 2014
ISBN 978-1-61451-603-3, e-ISBN (PDF) 978-1-61451-440-4,
e-ISBN (EPUB) 978-1-61451-915-7, Set-ISBN 978-1-61451-441-1

Speech and Automata in Health Care
Amy Neustein, 2014
ISBN 978-1-61451-709-2, e-ISBN (PDF) 978-1-61451-515-9,
e-ISBN (EPUB) 978-1-61451-960-7, Set-ISBN 978-1-61451-516-6

Technische Assistenzsysteme
Wolfgang Gerke, 2014
ISBN 978-3-11-034370-0, e-ISBN (PDF) 978-3-11-034371-7,
e-ISBN (EPUB) 978-3-11-039657-7

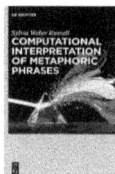

Computational Interpretation of Metaphoric Phrases
Sylvia Weber Russel, 2015
ISBN 978-1-5015-1065-6, e-ISBN (PDF) 978-1-5015-0217-0,
e-ISBN (EPUB) 978-1-5015-0219-4, Set-ISBN 978-1-5015-0218-7

Lukas König

Complex Behavior in Evolutionary Robotics

DE GRUYTER
OLDENBOURG

Mathematics Subject Classification 2010
68-02, 68M14, 68N19, 68N30, 68Q32, 68Q45, 68T05, 68T40, 68U20, 90C40, 93C85

Author
Dr. Lukas König
Karlsruhe Institute of Technology (KIT)
Institute of Applied Informatics and Formal
Description Methods (AIFB)
KIT-Campus Süd
76128 Karlsruhe, Germany
lukas.koenig@kit.edu

"Towards Complex Behavior in Evolutionary Robotics"
Von der Fakultät für Wirtschaftswissenschaften des Karlsruher Instituts für Technologie
genehmigte Dissertation
Datum der Prüfung: 25. Juni 2014
Referent: Prof. Dr. Hartmut Schmeck
Korreferent: Prof. Dr. Marius Zöllner

The author thanks Benjamin Bolland, Maximilian Heindl, Junyoung Jung, Serge Kernbach, Daniel Pathmaperuma and Holger Prothmann for kindly supplying photographs, and the Graphviz open source team for the free visualization software Graphviz.

ISBN 978-3-11-040854-6
e-ISBN (PDF) 978-3-11-040855-3
e-ISBN (ePUB) 978-3-11-040918-5
Set-ISBN 978-3-11-040917-8

Library of Congress Cataloging-in-Publication Data
A CIP catalog record for this book has been applied for at the Library of Congress.

Bibliographic information published by the Deutsche Nationalbibliothek
The Deutsche Nationalbibliothek lists this publication in the Deutsche Nationalbibliografie;
detailed bibliographic data are available on the Internet at http://dnb.dnb.de.

Cover illustration: Benjamin Bolland
Printing and binding: CPI books GmbH, Leck
♾ Printed on acid-free paper
Printed in Germany

www.degruyter.com

Acknowledgements

This book presents and concludes a major part of my research conducted from early 2007 until today. By far the greatest fraction of this time and, more importantly, the best of the presented ideas are strongly connected to the Institute AIFB where I worked as a research associate since December 2007. Here, my advisor Prof. Hartmut Schmeck gave me the opportunity to follow my research interests virtually limit-free in a creative and inspirational working environment, for which I am deeply grateful to him. I also thank him for his support and trust, even in difficult times, and for his constructive, detailed, often seemingly pedantic remarks which repeatedly led me to a better understanding of my own work.

However, as I will show in the book, any environment is inextricably linked with the agents acting in it, and this is particularly true of a working environment. Therefore, I am just as grateful to all my colleagues (former and current) for being so creative and inspirational, and for simply constituting an enjoyable company. In particular, I thank (in order of appearance) Sanaz Mostaghim, Ingo Paenke, Felix Vogel, Holger Prothmann, André Wiesner, Daniel Pathmaperuma, Christian Hirsch, Marc Mültin, Friederike Pfeiffer-Bohnen, Pradyumn Shukla, Micaela Wünsche, Fabian Rigoll, Fredy Rios and Marlon Braun for some of the most valuable (in very different ways) scientific discussions I ever had. Furthermore, I thank Ingo Mauser, Daniel Pathmaperuma, Friederike Pfeiffer-Bohnen and Felix Vogel for proofreading early, probably rather unpleasant versions of the manuscript.

Creating a doctoral thesis, including the many tasks it brings along, required the so far greatest effort of my life. This amount of work cannot be handled without being backed up in the "real world".

I thank my old friends, who never stopped being there for me, sometimes in person, always in my mind, and my new friends, the best of whom emerged from being mere colleagues at first, for reminding me of life beyond work. I am greatly thankful to my parents for encouraging me in not being satisfied with common opinions (which may well have become my most noticeable character trait today), to my grandfather for revealing me the joy of exploring everything and telling it to everybody (i. e., research and teaching) and to my grandmother for reminding me of the importance of studying (i. e., life beyond amusement). I adore my wife for showing me – a computer scientist – the beauty of nature (which may have given the thesis an esoteric touch at one place or another). Finally, I am most notably grateful to my whole family (naturally referring to my own as well as my wife's side) for supporting me in all those characteristic situations a PhD student gets into over time, and the few rather specific to myself.

Karlsruhe, December 2014   Lukas König

PS. Having little to do with the content of the book, I still cannot leave unmentioned an amazing contributing to a very special day: Thank you, little sister, for an exam celebration cake in the shape of my first computer – including an eatable *Basic* program.

Für Christine

Contents

List of Figures

List of Tables

List of Notations

1 Introduction

Preamble: *Nature has gained a huge variety of complex life forms which populate numerous biological niches all over the earth. The principles of natural selection, which are believed to be the main responsible forces driving the evolution of life on earth, seem to be a powerful mechanism capable of generating complexity in many different facets. But can artificial evolutionary approaches such as Evolutionary Robotics, which significantly simplify the biological mechanisms and work with population sizes and time periods incomparable to those in nature, also create truly complex structures? And can these structures, once evolved, be "reverse engineered" to provide an insight into their functioning? These are two major questions Evolutionary Robotics is dealing with today.*

Robotics is a well-established field of research continuing to yield robotic systems which are practically utilized in a wide variety of areas. Among these are industrial applications which typically use robots to perform tasks requiring great precision, speed, power or endurance (e. g., welding, painting, assembly); disaster control and military applications where robots are frequently deployed to do jobs that are hazardous to people (e. g., defusing bombs, exploring shipwrecks) or as unmanned (aerial) vehicles with various objectives; medical applications among which the most interesting are today located in the area of surgery assistance; and service or domestic applications where robots perform services to support people in everyday tasks (e. g., vacuum cleaning, wiping) or in special situations such as assistance in elder care. In many of these areas robots have already reached virtual perfection carrying out the task they have been designed for. This is facilitated by an extensive body of knowledge in the fields of electrical engineering, kinematics and mechanics, and by a growing experience in the design of robots for a number of specific applications.

However, there is an even bigger class of potential robot applications which are desirable and apparently realizable in principle, but which today still hold requirements that cannot be satisfied. Such requirements can be of a rather technical nature meaning, for example, that parts of a robot would have to be manufactured smaller (motors, sensors or the energy supply) or that the required motoric precision is greater than current technology is capable to provide. But beyond that, there are substantial unresolved issues concerning the programming of robots. For a great number of desired applications, today's algorithmic and software engineering techniques are insufficient to yield the according robotic behaviors. Especially when dealing with autonomous robots, even seemingly simple behaviors that are easy to describe on a high level in terms of "what a robot should do" can be hard to encode in a robot's programming language. This problem arises from the difficulty of capturing and interpreting correctly a robot's "world view", which is given by its sensory perception and inner state, and of foreseeing the different situations a robot might get into, particularly when interacting with other robots or living beings. Conversely, relatively simple behavioral code may cause robots to produce behavior which looks complex for an outside observer. Consequently, it is usually hard to decide from bare observation of a robot in specific situations what general type of behavior it is programmed to perform. This means that even when a behavioral program exists which (apparently) lets a robot perform a desired behavior correctly, it may be impossible to prove for that behavior that it actually has certain desired or undesired properties.

While this has long been known to be true from an algorithmic point of view (at least for any program written in a Turing-complete programming language), Valentino Braitenberg showed in 1984 that it can even hold for a single autonomous robot controlled by a program from a class of simple reactive functions, far from being Turing-complete [17]. Equipped with a few light or distance sensors and two wheels only, Braitenberg connected (mostly in thought experiments) the sensory inputs directly with the wheel motors in specific ways. In doing so, he was able to find examples of behaviors which, from the outside, give the impression of complex intentions or "feelings" such as love, aggression, foresight and optimism, but are in fact extremely simple to describe algorithmically. Of course, when knowing the underlying algorithms, such behaviors would not be considered complex, and on the other hand there exists a huge variety of complex behaviors for which no simple algorithms are known today.

In a *robot swarm*, i. e., a collection of at least two (usually many) interacting autonomous robots, the discrepancy between an observed or desired behavior and the according algorithmic description becomes even greater. There, the correct interpretation of sensory information, which is usually rapidly changing and impaired by interfering sensors of other robots, as well as the understanding of swarm dynamics on a theoretical level still involve many unresolved problems. A classic engineering approach for creating programs for a swarm of robots might involve (1) describing the swarm behavior as the result of interactions among individual behaviors, and (2) encoding the individual behaviors into controllers. While the second step is essentially the above-discussed problem of programming single robots, the first step contains additional complexity in decomposing a process that is a result of dynamical interactions among its sub-components. Altogether, it seems hard to imagine a general approach, suitable for a large class of desired behaviors, to be based on this or any other engineering process known so far.

On the other hand, desired applications for autonomous robots have inspired the fantasy of science fiction writers for a long time, and, used wisely, such technology could improve the lives of many people in various ways. Applications range from rather simple tasks such as cleaning and other home service purposes or Braitenberg's behaviors mentioned above to sophisticated tasks such as searching for objects with specific properties in an unknown and complex environment, exploring unknown areas in a group by creating and exchanging map information, or recognizing and manipulating objects which are arbitrarily located and oriented in space in an elaborate way. These rather technical behaviors are essential for ambitious applications such as, for example, searching and rescuing people in a disaster area, or the even more challenging application of floating through a human body with the aim of curing diseases. An additionally aggravating property of these and many other applications is that they are implicitly *decentralized*. This means that the robots have no or little help from "outside", such as a central computer providing them with maps, their current position etc. Instead, they have to base all their decisions on local observations and within-swarm communication only.

A common approach for finding decentralized strategies suitable to solve tasks in a robot swarm is to observe swarm-like populations in nature [113]. For example, populations of ants or bees have evolved adaptive behaviors emerging from simple local interactions with their environment and each other, and they are usually close to solving their biological challenges in a nearly optimal way. By a careful analysis of the behavior of natural swarms, local rules for collective behavior can be extracted in some cases [14]. One of the most prominent examples of successful behavior extraction from nature is the well-known field of *Ant Colony Optimization* which is based on (a rigorously abstracted version of) the capability of ants to find shortest paths from a nest to a foraging place [57]. However, there is no guarantee that the behavior of a natural swarm can be understood sufficiently well to successfully transfer the according rules to an artificial swarm. Moreover, there are many desirable tasks for artificial swarms which do not have a related counterpart in nature. Additionally, generating programs manually or by imitating the behavior of living beings is challenging and expensive, and it can be infeasible in many cases.

It is not foreseeable if robot behaviors as complex as the ones described above will some day be programmable manually using a structured engineering process. Then again, in complex or dynamic environments it might even be disadvantageous to provide a robot with a monolithic, fixed program. In this context, it seems reasonable to introduce a flexible and adaptive component into a robot's behavior which lets the robot learn while exploring an environment and base decisions on past observations. In the last two decades various approaches have been proposed which provide robots with a high-level description of the task they are supposed to perform, omitting a detailed description of the environment or of how exactly to perform the task. The robot, in turn, has to figure out a way of solving the task in the environment it is placed into. In some sense the robot is supposed to program itself on the low level of atomic actions while comparing the results to the high-level description of a desired behavior.

This concept is followed in the fields of *Evolutionary Robotics (ER)* and *Evolutionary Swarm Robotics (ESR)*. Using mechanisms inspired by natural evolution, ER and ESR in a sense generalize the above-mentioned principle of observing specific biological populations to mimicking the overall process of "problem solving" in nature. The potentials of this idea are evident in any living animal organism, be it as simple as a fly or a spider – not to mention higher animals and humans. These "natural robots" emerged from evolution to be capable of performing highly complex decisions and actions following their desire of living fertile – or even "happily" – to adulthood. All

of them are incomparably more complex than any artificial robot to be produced anytime in a foreseeable future. However, before real advantage can be taken from this
powerful principle, there is still a knowledge gap to close which in its essence can be
summarized as the following question: (how) can the complexity of natural evolution
be sufficiently abstracted to provide a technology feasible for human roboticists by
still being capable of evolving sophisticated robotic systems?

1.1 Evolutionary Robotics and Evolutionary Swarm Robotics

The basic ideas underlying ER have originally been borrowed from the fields of *Artificial Intelligence (AI)* [168] and *Evolutionary Computation (EC)* [35, 77, 111]. However,
one of the first researchers who thought of an evolutionary process driving the generation of control systems of artificial organisms was Alan Turing who suggested this
idea as early as 1950 [193]. In the years 1992 and 1993, the first experiments on artificial
evolution of autonomous robots were reported by a team at the Swiss Federal Institute of Technology in Lausanne, a team at the University of Sussex at Brighton, and
a team at the University of Southern California. The success and potentials of these
studies triggered a whole new activity in labs across Europe, Japan, and the United
States. Since 1993 these techniques have been summarized in the field of ER [28, 142],
which is considered to roughly capture all techniques for the automatic creation and
programming of autonomous robots which are inspired by the Darwinian principle of
selective reproduction of the fittest [32].

In ER, robots are considered to be autonomous organisms that develop a control
system on their own in close interaction with their environment and usually without
human intervention. Some ER approaches also include the evolution of a body configuration for robots, in addition to or instead of a behavioral evolution. The ER methods
share important characteristics with biological systems, above all robustness, flexibility and modularity. Moreover, a benefit of ER over classic methods is its rather low requirement of technical skills meaning that it can be utilized and controlled by people
with different knowledge backgrounds. Today, the ER community ranges over different fields of research including AI and robotics as well as biology and cognitive science
and the study of social behavior [46, 47, 142].

An actual evolutionary process for robot behavior is usually designed as a mixture
of classic EC components and further nature-inspired components accounting for the
embodiment of the individuals to evolve. Overall, a collection of random or predefined
controllers is given as an initial population which is repeatedly mutated and recombined while selecting subpopulations for reproduction, considering a rating of their
ability to perform a desired task (*fitness*), and discarding the others. The fitness of the
controllers is calculated by some observer which decides how close the performed
behavior matches the desired behavior; this can be the onboard computer of a robot

itself (in *decentralized* approaches such as the ones suggested in this book), a distinct outside computer or even a human.

A rather new subfield of ER is ESR which comprises those approaches from ER that involve the evolution of self-organizing group behaviors of several individual autonomous robots. The term has been coined quite recently by Vito Trianni in 2008 [190]. In ESR, evolutionary techniques from ER are used in order to obtain robust and efficient group behaviors based on self-organization. A more theoretical challenge concerns the understanding of the basic principles underlying self-organizing behaviors and collective intelligence. Besides ER, the fields of swarm intelligence, swarm robotics and multi-agent systems are closely related to ESR.

It is still an important challenge of ESR and robotics in general to design purposeful and complex swarm behavior. Such behavior should be technically useful, adaptive to environmental changes and scalable in size and functional metrics [30, 110]. The application of evolutionary techniques to swarm robotics has been known before the term ESR appeared, cf., for example, [47, 115, 122, 206]. Besides the evolution of complexity, essential challenges for a successful utilization of evolutionary approaches in ESR include decentralization as well as obtaining all evolutionary results during one life-cycle of a robot (*online evolution*). These and two further classifications of evolutionary processes in ER and ESR are discussed in the next section.

1.2 Further Classifications

Approaches in the area of ER as well as ESR are commonly classified along four dimensions:
1. simulation – reality,
2. offline – online,
3. offboard – onboard (decentralized), and
4. encapsulated – distributed.

The first distinction (simulation – reality) addresses the choice of the experimentation platform for ER research. Experiments in ER are often carried out in simulation as the evolution process may require multiple trials and a large amount of time as well as costly robot hardware that can be damaged and is subject to wear [176]. Results from simulation, however, often cannot be readily transferred into reality which is known as the *reality gap* [127]. Therefore, it is believed to be necessary to work at least partially with real robots [199]. A fairly reasonable compromise can be to obtain a mixture of studying the basic evolutionary principles in simulation and adapting the parameters in detail using real-robot experiments.

The second axis (offline – online) focuses on the time at which evolution takes place. If controllers are evolved in an artificial training environment (simulation or laboratory experiments) before the execution of the outcoming controllers in the actual area of interest, it is called an offline process. Due to deficiencies of the training environment, offline methods have to deal with a smaller version of the reality gap, too. In contrast, online methods are based on evolution which takes place during execution of the actual task in a real environment. Therefore, the robots are supposed to evolve and solve the desired task simultaneously (or, at any rate, first to evolve while not executing any harmful behavior, and then to solve the task using the evolved controllers). Online approaches can be tested in artificial environments, but the resulting controllers are not intended to be used in a real environment as they are not expected to accurately match the real-world requirements. This approach has the advantage of the real environment being involved in the learning process which often cannot be modeled precisely beforehand. On the other hand, a problem of online approaches is that the uncertainty of success, which is present in every learning approach, can have more fatal consequences when real hardware in a real environment is involved. Therefore, important research questions involve the reduction of the reality gap and the prediction of success before an actual run in a real environment. These questions are addressed in Chap. 6 by proposing a formal framework for success prediction of evolutionary processes involving complex environments.

The third distinction (offboard – onboard or decentralized) points at the presence or absence of a central computing unit during evolution. Onboard methods, also called decentralized methods, involve approaches where the onboard computers of the robots carry out all behavioral decisions themselves or by communicating only with "neighboring" robots of the swarm (the term "neighboring" can be defined in various ways, however, often it is simply given by the communication radius of the robots). Conversely, offboard methods use a central computing unit which observes and controls the evolution process to some amount from "outside". For example, robots equipped with a GPS system which provides information about their position on the earth surface would be considered partially offboard-controlled. With a decentralized approach, only local sensor information and internal resources such as memory and processing power can be used. This may seem as a limitation at first sight, but it offers some important advantages as well:

- *Scalability*: With a rising number of robots in the population, any kind of a central resource eventually may become a bottleneck in terms of computing or communication power (even a GPS system, when using billions of robots in a future scenario; however, offboard approaches today usually depend on a much less robust computing and communication system than GPS, by often providing sophisticated information to the evolution process such as which robots to select for reproduction).
- *Robustness*: The central resource provides a single point of failure which the whole evolutionary process depends on.
- *Feasibility*: The presence of a central unit is just infeasible in many types of real environments.

The fourth axis (encapsulated – distributed) divides serial from parallel methods, encapsulated approaches being the more serial ones. In an encapsulated approach, all individuals of a population are hosted on a single robot and get evaluated sequentially. In contrast, in distributed approaches every robot hosts only one individual, the population consists of several robots and the fitness evaluation takes place in parallel. The latter variation is closely associated with the decentralized approaches and offers similar advantages, such as robustness and scalability. In encapsulated approaches, the time needed in order to evaluate the entire population increases linearly with the number of individuals. By using parallelization, evaluation time can be reduced up to essentially constant time in the distributed case.

The models and experiments presented in this thesis are positioned at the "online (2) / onboard (3) / distributed (4)" section of this classification space. Regarding the first axis, many experiments have been performed in simulation, while the fundamental results of the second main research part of the thesis (Chap. 4) have been verified with real-robot experiments. The experiments from the third part (Chap. 5) have been performed exclusively in simulation due to the great amount of program memory needed for the implementation of the "completely evolvable genotype-phenotype mapping" which exceeded the memory present on the "Wanda" real-robot platform (cf. Sec. 4.4). The experiments of the fourth main research part (Chap. 6) have also been performed in simulation only, as the implementation on Wanda has been beyond the scope of this thesis. Their implications concerning the reality gap therefore have to be considered preliminary. However, overall it is a major goal of this thesis to illuminate basic principles of evolution in robot swarms. These principles are much less dependent of the underlying platform (simulation or robotic) than, e. g., individual robot behaviors are [199] and can, to a great extent, be generalized from simulation studies to real-robot platforms. A transfer from simulation to reality can, thus, be considered merely a parameter adjustment, as is shown in Sec. 4.4 for the evolutionary approach presented in Chap. 4.

1.3 Challenges of ER

The field of ER has been established for a bit over two decades now, and it has had some major breakthroughs in generating behaviors automatically. Particularly, various behaviors have been evolved which are, up to today, hardly programmable by hand, for example [5, 37, 42, 92, 138, 190, 191]. Moreover, ER has proven to be capable of finding control mechanisms which outperform manually designed controllers in terms of effectiveness in solving a desired task and compactness of the controller [199]. However, there are still drawbacks that current ER research has not been able to eliminate. Two of the most important ones concern
- the evolution of complexity: so far, it has not been possible to show that truly "complex" behavior can be evolved by purely evolutionary approaches [139]; nev-

ertheless, there has been progress toward the evolution of complexity by means of mixing evolution with other techniques or types of world knowledge, for example hierarchical evolution [38]; and

– the assurance of correctness: in most cases the programs generated by evolution cannot be analyzed automatically to prove that the behavior they produce is in fact within a range of desired behaviors.

Both drawbacks have serious consequences. The first questions the usefulness of the field as a whole as the main purpose of ER is to produce behaviors of a certain complexity that cannot be programmed by hand. Even for the most complex behaviors evolved so far it seems perfectly possible to imagine some (future) engineering process by which they could be generated in a structured way, requiring moderate effort. Although ER approaches still may have benefits such as being more efficient in terms of man-hours or compactness of the code, or as being relatively simple to use, it is still unsatisfactory that none of the evolved behaviors significantly outperforms human solutions in terms of complexity.

The second drawback points at the usefulness of the resulting behaviors. While, by pure observation, a behavior may seem to be carried out correctly, it is unclear if this holds for all imaginable situations the robots can be in. This uncertainty can potentially lead to fatal consequences for involved robots, materials or humans. The well-known software engineering rule that testing can never assure a program's correctness is applicable even more strictly to programs for autonomous robots as it can be hard to decide even for specific test cases if they perform correctly. Moreover, up to today most ER approaches use the classic *Artificial Neural Network (ANN)* model as a control structure which is known to be hard to analyze. In their most general version, ANNs are Turing-complete [82] which means that no semantic properties can be proven for them at all in an automatic way (according to Rice's theorem [165]). For such control structures, behavior observations are generally the only way to tell what a robot or a swarm is doing. On the other hand, for many desired robot applications it is crucial to know for certain that the robots perform correctly at any time. For example, regarding a medical application a promise such as "we never saw it behave harmfully" will not suffice for a robotic system to be introduced in a hospital.

As most current research is focussed around the first drawback, i. e., the evolution of complexity, there is only little work concerning the analysis of evolved behaviors. This thesis, beyond studying the former, addresses the latter by using a *Finite State Machine (FSM)*-based control structure for robot behaviors. This choice has been made in favor of the analyzability of evolved controllers and by deliberately sacrificing the higher descriptive power and simpler training methods of ANNs. Throughout the thesis, several examples are given of how controllers can be analyzed automatically for non-trivial behavioral properties. Furthermore, the experimental studies in Chaps. 4 and 5 show that the lack of descriptive power and simple training methods of FSMs can be overcome by appropriate evolutionary operators and settings, at least as far as behaviors of complexities comparable to current literature are concerned (counting only purely evolutionary approaches). As behavioral complexity rises, the advantages of ANNs over FSMs may gain importance, however, such a level of complexity does not appear to come into reach for ER anytime soon.

1.4 Structure and Major Contributions of the Thesis

This thesis is focussed around research in the field of ESR, and, more generally, the fields of ER and evolution in complex environments. Within these fields, two major topics are

1. the evolution of robot behavior using FSMs as control structures, for which correctness can be automatically verified (Chap. 4), and
2. the influence of a non-trivial (at its highest stage itself evolvable) transformation from a genotypic representation of robot behavior to the behavior itself, on the evolution of complexity (Chap. 5).

Another major topic of the thesis, accentuating a more high-level focus, is

3. the prediction of success before an actual evolutionary run is performed, with the aim of avoiding expensive failures in experiments with real robots (Chap. 6).

Furthermore, the field of *Agent-based Simulation (ABS)* is studied along the way as large parts of the evolution experiments have been carried out in simulation. For this purpose another major part of the thesis has been

4. the creation of an ABS framework that has specific advantages over existing frameworks – particularly including an implicit structural programming guidance for unskilled users [109] that can be used as an architectural pattern in other simulations, too (as this framework is inevitable to provide for reproducible experimental results, it is described first within the main research part of the thesis, i. e., in Chap. 3).

The ABS framework introduced in Chap. 3 is called *Easy Agent Simulation (EAS)*, and it has been used for all simulation experiments presented in this thesis. The real-robot experiments have been performed on a robot platform called *Wanda* that has been developed in cooperation with the *Institute for Process Control and Robotics (IPR)* at the *Karlsruhe Institute of Technology (KIT)* [96], cf. Chap. 4.

According to the focus structure above (cf. also Fig. 1.1), the major contributions of the thesis can be stated as follows:

1. Proposal and comprehensive evaluation of a novel FSM-based control model for robots and a corresponding decentralized online-evolutionary framework, both applicable to various types of robots. Results show that the framework is capable

of evolving robot behaviors of complexities comparable to current state-of-the-art approaches from the literature, by yielding fairly analyzable robot programs.

2. Proposal and evaluation of a highly flexible *Genotype-Phenotype Mapping (GPM)* which, by mimicking the process of DNA to protein translation in nature, can be evolved along with robot behavior, leading to an automatic adaptation of the evolutionary operators to a given search space structure. Results show that the approach can lead to such an adaptation as intended and furthermore to a significant improvement of the evolutionary outcomes. Nevertheless, due to a multitude of influences and adjustable parameters, it has been beyond the scope of this thesis to conclusively clarify, to which amount the improvement can be ascribed to the flexibility of the GPM.

3. Proposal and evaluation of a formal framework for the prediction of success in evolutionary approaches involving a complex environment. By measuring the qualities of both fitness-based and environmental selection in variable levels of abstractions from the real scenario (purely probabilistic estimations, simulation runs, laboratory experiments etc.), the probability of a successful behavior evolution in a real environment can be predicted. Results show that the model is capable of predicting the success of (so far rather abstract) simulation scenarios with a high precision.

4. Proposal, implementation and evaluation of a programming pattern for agent-based simulations intended to improve structuredness and code reusability in implementations by non-expert programmers. Experiments with six student test persons suggest that the proposed architecture implemented in EAS yields more structured and reusable implementations than the state-of-the-art simulation programs MASON and NetLogo. The framework is available including all sources and documentation at sourceforge [101].

Many of the results presented in this thesis have been published before, mostly as peer-reviewed articles in conference proceedings or journals. These publications provide a contentual foundation of the thesis, but the thesis content is not limited to the sum of their content. The text of the thesis has been written either from scratch or by majorly rephrasing passages from existing own publications, providing in many cases more comprehensive descriptions, deeper formal analyses and more thorough discussions. Furthermore, several formerly unpublished results have been included in the thesis. Details about the publication history and the novel findings presented are given in each of the main research chapters.

The overall thesis structure is depicted in Fig. 1.1. Following this introductory chapter, the chapter entitled "Robotics, Evolution and Simulation" summarizes related work in the relevant fields of research and gives definitions of the fundamental concepts, models and terms used throughout the thesis. The four subsequent chapters build up the main research part of the thesis (painted gray in the figure). Chap. 3, entitled "The Easy Agent Simulation", introduces and discusses the simulation platform

Chapter	Topic
1	Introduction
2	Robotics, Evolution and Simulation
3	The Easy Agent Simulation
4	Evolution Using Finite State Machines
5	Evolution and the Genotype-Phenotype Mapping
6	Success Prediction of Evolution in Complex Environments
7	Conclusion

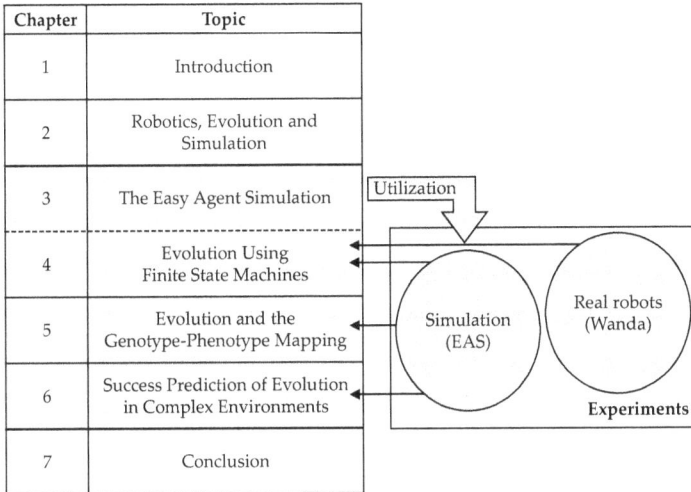

Fig. 1.1. Structure of the thesis. The table on the left depicts the structure on a chapter level. A gray background indicates the main research part of the thesis. Research concerning the simulation plugin interface led to a simulation framework called EAS. The box on the right indicates how the theoretical work concerning robot evolution is experimentally evaluated using EAS and the robot platform Wanda.

EAS used for the ESR experiments. The *Simulation Plugin Interface (SPI)* is proposed as an architectural pattern, and its implementation in the EAS Framework is discussed and evaluated in a comparative study. Chap. 4, "Evolution Using Finite State Machines", introduces an FSM-based control structure for robots and a corresponding online-evolutionary model including mutation, recombination and selection operators. The model is evaluated using the EAS Framework and the robot platform Wanda. Chap. 5, entitled "Evolution and the Genotype-Phenotype Mapping", introduces the concept of a *(completely) evolvable GPM* meaning that the mapping defining the translation of genotypes into phenotypes, i. e., robot controllers, is subject to mutation during evolution just as the robot controllers are. The model is evaluated using the EAS Framework. In Chap. 6, entitled "Data Driven Success Prediction of Evolution in Complex Environments", a mathematical framework based on Markov chains is proposed which is intended to calculate the expected success of evolution, particularly in ESR. As an application, the online and onboard-evolutionary approach presented in this thesis is used to evaluate the model in simulation utilizing the EAS Framework. Finally, the chapter entitled "Conclusion" gives a summary of the results presented in the thesis and provides suggestions for further research.

Throughout the thesis, a short abstract captioned "Preamble" precedes each chapter, providing an informal comment on the subjects to be treated. Rather than systematically summarizing a chapter, the abstract's purpose is to classify a chapter's most important idea or research question with regard to its implications within the scientific landscape and "the real world" — as perceived by the author.

2 Robotics, Evolution and Simulation

Preamble: *When watching a swarm of autonomous robots perform even the simplest actions such as driving around randomly while avoiding to crash into each other, an impression of a highly complex and unpredictable system arises, and it is hard not to draw an analogy to populations of living organisms. Somehow it seems obvious to try to transfer some of the intriguing mechanisms from nature to such artificial robot populations and to look if an intelligent system can emerge. Trying to do so, it soon gets obvious that there is still a long road to follow until the underlying mechanisms will be understood sufficiently well to be used in a practical way for the automatic generation of intelligent, robust and controllable robot behavior. But no matter how long it takes, the idea never seems to lose its fascination.*

Undoubtedly, there has been great progress in the fields of robotics, artificial evolution and simulation in the last decades. Each field individually as well as interdisciplinary fields such as ER and multi-agent systems have evolved and gained new insights to a large extent. This chapter summarizes the current state of research related to this thesis, introduces basic terms, preliminaries and concepts, and gives a general motivation. First, learning and particularly evolutionary techniques are introduced by focussing on the fields of EC, ER and ESR. Particularly, the roles of the controller representation in ER and of the GPM in general are discussed. In the second part of the chapter, current research in the field of simulation is presented. Each of these fields has a huge list of publications, therefore, only the most relevant work can be outlined and referenced here. The numerous significant articles going beyond the scope of this thesis can be found, for example, in several survey papers referred to below. Particularly, the works by Meyer et al. [127], Walker et al. [199], Jin [86] and Nelson et al. [139] summarize the fields of ER and ESR, the work by Floreano et al. [46] deals with *neuro-evolution* (i. e., evolution of ANNs) in ER, and the works by Railsback et al. [157] and Nikolai and Madey [140] give a survey of general-purpose ABS.

2.1 Evolutionary Training of Robot Controllers

ER is a methodology for the automatic creation of robotic controllers. ER approaches usually borrow the basic ideas from classic EC approaches which, in turn, model highly abstract versions of the processes which are believed to be the driving factors of natural evolution. Classic EC methods are meta-heuristics, usually used to solve computationally hard optimization problems [201]. Beyond that, EC has been successfully used to evolve programs for the calculation of specific mathematical functions [111], the shapes of workpieces such as pipes [161] or antennas [81], for the optimiza-

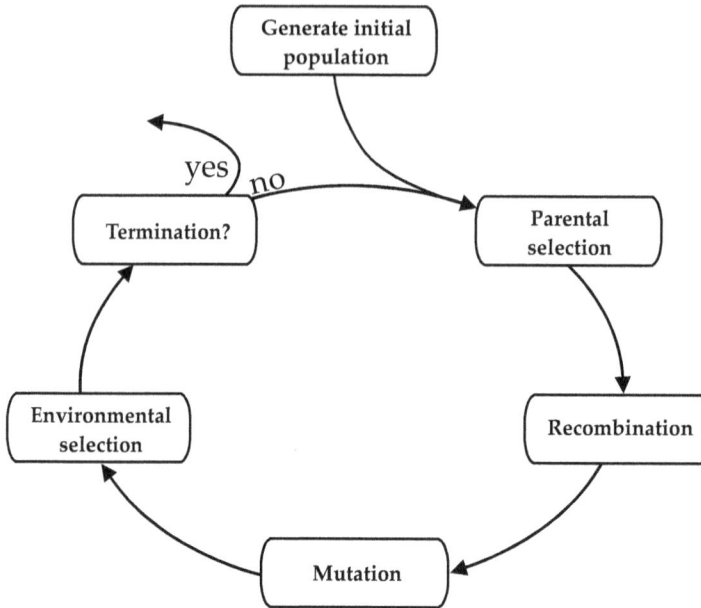

Fig. 2.1. Classic evolutionary cycle as used in many EC approaches. The process starts with a – usually randomly generated – initial population of solution candidates for some (optimization) problem. The candidates are rated using a fitness function which is usually given more or less directly by the problem definition. Based on the rating, a "parental" selection is used to create a sub-population of expectedly superior individuals from which (optionally by using a recombination operation) an offspring population is created. The individuals from this population are subsequently mutated and, optionally, subjected to an additional "environmental" selection. If a termination criterion is reached, the algorithm stops, otherwise the next iteration of the loop is started.

tion of non-functional properties of programs [205], rather playfully for the generation of specific "magic squares" [161], and for various other purposes. Classically, EC has been composed of sub-fields such as Evolutionary Algorithms, Evolution Strategies, and Genetic Algorithms. Today this differentiation is not considered to be strict anymore by most authors, and it does not affect the results presented here. Therefore, it is not further attended for the remainder of the thesis, and, rather, the term EC is used.

Most EC approaches have in common to use the basic evolutionary cycle, depicted in Fig. 2.1, or a modified version leaving out, for example, one of the two selection operations, the recombination operation (cf., for example, the Evolution Strategies [160]), or even the termination condition in open-ended evolutionary approaches [158].

Similar to the procedure in classic EC, populations of individuals in ER evolve to achieve desired properties by being subject to selection, recombination, and mutation. There, individuals represent robot controllers, more high-level behavioral strategies for swarms or even robot morphologies [132]. This process can be performed with a

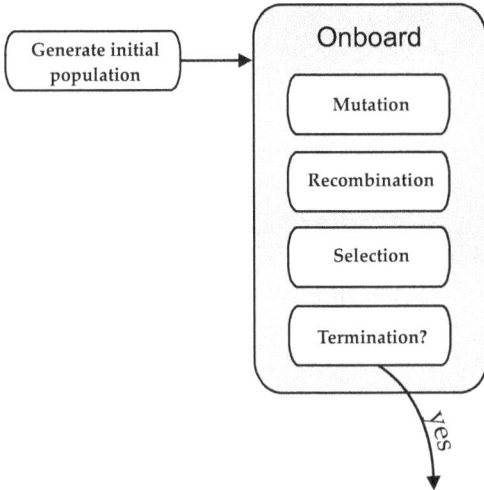

Fig. 2.2. Evolutionary process onboard of a robot. In a decentralized ER scenario, the evolutionary operators mutation, recombination and selection are performed in a local way on every individual robot or using local robot-to-robot communication.

single robot evaluating each individual of the population successively (encapsulated), or a population of robots, each storing and evaluating one individual (distributed). In contrast to EC, fitness usually has to be measured by observing robot behavior in an environment rather than being calculated directly from the individual. Furthermore, the environment itself provides for part of the selection pressure implicitly which affects evolution apart from the influence of fitness (cf. below and Chap. 6). Overall, the subparts of the course of evolution in ER can take place in an asynchronous way which is why a more parallel view is better suiting for ER than the cycle view of classic EC. Particularly when performing decentralized evolution, the whole evolution process takes place onboard of robots letting mutation, recombination and selection of individual robots occur in parallel. In that case, particularly selection cannot be modeled as a global operator, but has to be performed in a local way through robot to robot communication. Fig. 2.2 depicts the basic evolutionary process taking place onboard of a robot.

While the general evolutionary process described so far can be used to evolve robot behaviors as well as robot morphologies, the remainder of the thesis focuses on the evolution of robot behaviors. Various types of robot behaviors have been evolved in the past, the most popular of which has been *Collision Avoidance (CA)*[1] (or *Obstacle Avoidance*). CA denotes behaviors including a capability of driving (or walking etc.) around while avoiding to crash into obstacles or other robots [139]. Other behaviors include *Wall Following* [184], *Object Detection* and *Recognition* [121, 207], *Shortest Path Discovery* [185], *Object Transport* [59], and others; the references denote exemplary examples out of bigger pools of publications referring to the respective behaviors. Beyond these behaviors which can be performed by a single robot or in parallel by a population of robots, a popular behavior specifically suited for swarms of robots is *Group Transport* [61] which denotes the collective transport of an object usually too heavy for a single robot to carry or push. Other popular swarm behaviors include *Coordinated Motion* [37, 167], *Bridge Traversal* and *Gap Crossing* [25], and others.

2.1.1 Two Views on Selection in ER and ESR

Evolutionary approaches in the field of ER and its subfield ESR tend to be more closely related to natural evolution than classic EC approaches, as the former include physical individuals that operate in a spatial environment and are confronted with physical constraints as well as inherent benefits such as parallelism. The main differences between the fields are focussed around selection and fitness calculation as these operations usually include a strong interaction with a complex environment in ER, as opposed to most classic EC approaches. In this respect, ER can be looked at from two different perspectives, namely a classic EC perspective and a biological perspective, both of which can yield specific insights.

The classic EC point of view. In classic EC, selection is usually performed by considering the "fitness" of all individuals of a population and favoring the "better" ones [201]. There, fitness is given by one or several numeric value(s) calculated directly from an individual, and it reflects the individual's relative quality, often with respect to an optimization problem to solve. In contrast, fitness in ER is computed by observing an individual's performance in an environment. This environmental fitness calculation can be noisy, fuzzy or time-delayed [18, 75, 151] which is why, at some places in this thesis, it will be considered a "measurement" rather than a calculation. Obviously, environmental circumstances can highly affect the success of an individual in terms of its ability to collect fitness points. Additionally, the environment is in many cases responsible for an implicit pre-selection of individuals. In such cases the individuals have to match some environmental requirement to even be considered for fitness-based selection. For example, a common practice in a decentralized swarm

---

1 For example, the following publications describe experiments regarding the evolution of CA on real robots with different locomotive capabilities (possibly among other behaviors): [7, 36, 40, 44, 45, 66, 74, 75, 84, 97, 118, 121, 124, 128, 136, 137, 138, 143, 145, 148, 149, 188, 208, 209]

scenario is to let robots perform reproduction (including selection) when they meet, i. e., come spatially close to each other. Then, evolution implicitly selects for the ability to find other robots – in addition to selection based on the fitness value [19]. Nolfi and Floreano refer to these two different factors influencing selection as *explicit* vs. *implicit fitness*[142], the former denoting a calculated (or "measured") fitness value, and the letter a robot's quality in terms of the environment-based part of selection. In the following, these terms are adopted, using the term *fitness* as an abbreviating alternative when explicit fitness is meant. Overall, a complex environment adds fuzziness, noise and time-dependencies to the calculation of fitness and it introduces an implicit factor to the selection process, compared to a classic EC approach.

The biological point of view. In evolutionary biology, an individual's *reproductive fitness* is calculated from its ability to both survive to adulthood and reproduce with an expectation of contributing to the gene pool of future generations [52]. There are several competing approaches on how to exactly calculate reproductive fitness in nature (particularly short-term vs. long-term calculations [175] etc.). In any case, reproductive fitness in nature as well as explicit and implicit fitness in ER reflect the expectation of an individual to produce offspring genetically related to itself. However, reproductive fitness in natural evolution is an observable but (mostly) unchangeable property. Particularly, natural evolution by itself does not have an explicit target which it aims to, but its direction is solely dictated by interactions of individuals with their environment, and by the selection pressure established by these means. In contrast, explicit as well as (to a lesser extent) implicit fitness in ER can be designed to guide evolution in a certain direction. There, explicit fitness is often rather straight-forward to design (although possibly complicated to adjust) as, by the basic idea of ER, it captures properties that can be more easily encoded on a high level than programmed on a behavioral level. For example, when evolving CA, driving can yield positive fitness points while being close to a wall or colliding can be punished. Implicit fitness, on the other hand, is more difficult to influence as the entire environment has to be designed accordingly. Furthermore, implicit fitness can include complex long-term properties such as, e. g., a robot promoting its own offspring in producing new offspring. This is due to the fact that offspring behavior contains partially the same genes (i. e., pieces of behavioral descriptions) as the according parental behaviors. Therefore, implicit fitness of the parent robots can increase indirectly when the offspring chance of contributing to the gene pool increases.

As opposed to nature, evolution in ER is supposed to be controlled to move into a certain direction. Therefore, the explicit fitness function and, to some extent, implicit environmental selection properties can be designed according to desired behavioral criteria. To give another example, the following situation can be considered: let a swarm of robots be explicitly selected for the ability to find a shortest path from a nest to a forage place, for example, by giving a fitness bonus reciprocally proportional to the time required for the travel between nest and forage place. As long as

Table 2.1. Classification of fitness functions in ER. The table lists seven types of fitness functions one of which most ER approaches can be assigned to. The second column shows the a priori knowledge required for the use of the according fitness function type (the less a priori knowledge required the more flexible the learning process).

Fitness function class	A priori knowledge incorporated
Training data fitness functions (to use with training data sets)	Very high
Behavioral fitness functions	High
Functional incremental fitness functions	Moderate – high
Tailored fitness functions	Moderate
Environmental incremental fitness functions	Moderate
Competitive and co-competitive selection	Very low – moderate
Aggregate fitness functions	Very low

the robot density is sufficiently low to allow arbitrarily many robots to simultaneously traverse each path in the environment while keeping the interference of robots with each other low, explicit fitness will be the strongest selective power, hopefully leading eventually to the discovery of the shortest path. If, however, the shortest path is too narrow to fit more than a few individuals at a time, evolution might implicitly select for individuals that use longer, but less crowded paths, or for those that can decide which path to take based on congestion rates. Depending on the exact properties of the different paths, implicit selection might highly affect evolution in this example. In [19], the impact of environmental selection has been experimentally investigated in an even more extreme scenario using mainly implicit selection with a minor explicit part (cf. Sec. 6.4.3 of Chap. 6). There, robots learned to explore the environment by being selected for mating when they came spatially close to each other. In a second experiment, robots learned foraging by being implicitly forced to collect energy or to die otherwise. Overall, it turns out that in complex environments explicit fitness can play a subordinated role with implicit fitness having the major impact on selection. On the other hand, explicit fitness is easier to design in a proper way to drive evolution into a desired direction. Therefore, it is desired to arrange evolutionary settings such that major selection pressure is induced explicitly with implicit factors playing a supportive role.

2.1.2 Classification of Fitness Functions in ER

According to Nelson et al. [139], fitness functions in ER (including both explicit and implicit aspects) can be classified along the scheme depicted in Tab. 2.1. The basis for this classification scheme is the degree of a priori knowledge that is reflected in the fitness functions. A priori knowledge is knowledge about the system that is not evolved but hard-coded by humans before the beginning of an evolutionary run. While a priori knowledge can facilitate achieving a desired behavior in a certain environment, it makes the process less generalizable to arbitrary environments. For example, a fitness function with the purpose of rating a maze escaping behavior might involve fitness bonuses for crossing certain check points on the way out. As such check points are specific to the maze to be solved, they do not easily generalize to other mazes making the approach somewhat static.

The reason for choosing this classification scheme is that a priori knowledge reflects the level of truly novel learning that has been accomplished [89]. The more a priori knowledge is put into a learning process, the less true learning has to be accomplished by the system. In fact, from an abstract point of view a regular static behavioral algorithm can be considered a special case of a learning algorithm with so much a priori knowledge that no flexibility is left to the learning process at all. Fitness functions that require little a priori knowledge, on the other hand, can be used to exploit the full richness of a solution space, beneficial aspects of which may not have been evident to the experimenter beforehand [37, 142, 190]. The types of fitness functions listed in the table are outlined in the following according to the definitions given by Nelson et al.

Training data fitness functions. Fitness functions of this type require the most a priori knowledge. They can be used in gradient descent optimization and learning scenarios, for example for training ANNs by using error back propagation. There, fitness is directly connected to the error that is produced when the trained controller is presented a given set of inputs with a known set of optimal associated outputs, fitness being maximized when the output error is minimal. A training data set has to contain sufficient examples in order that the learning system can extrapolate a valid generalizable control law. Thus, an ideal training data set (which is in ER in most cases unavailable and infeasible to create) contains knowledge of all salient features of the control problem in question. Several examples of fitness functions of this type being successfully used in learning scenarios are given by Nelson et al. [139].

Behavioral fitness functions. Fitness functions of this type are typically rather easy to assemble which is why they are widely used in ER. Behavioral fitness functions are task-specific and hand-formulated, and they can measure various aspects of what a robot is doing and how it is doing it. In distinction to more elaborate fitness function types, behavioral fitness functions measure how a robot is behaving, but typically not what it has accomplished with respect to the desired behavior. For example, a behavioral fitness function for CA could observe the wheel motors and proximity sensors of a robot and initiate a bonus when the robot drives forward while punishing it when approaching an obstacle. However, this fitness calculation does not consider the robot's success on a task level which is given by moving crash-free and preferably quickly through a field of obstacles. Such fitness calculation includes a priori knowledge of how a robot should learn a desired behavior, which introduces the risk of cutting out other possible solutions. For example, in the case of CA, communicating with other robots to get information about crowded places might lead to improved results, but this solution is precluded when using the above fitness function. In contrast, aggregate fitness functions measure aspects of what a robot has accomplished, without regard to how it has been accomplished (see below). There are many examples of behavioral fitness functions in the literature, e. g., [7, 84, 118]. In this thesis, the fitness function for CA is a behavioral fitness function.

Functional incremental fitness functions. This type of fitness functions is used to guide evolution gradually from simple to more complex behaviors. The evolutionary process begins by selecting for a simple ability such as, for example, CA. Once the simple ability has been evolved, the fitness calculation method is replaced by an augmented version which induces selection for a more complex behavior. For example, a task of maze escaping might be activated after a foundation of CA has been evolved. In an evolutionary run, the fitness calculation can be incremented in several stages of increasing complexity. When evolving complex behaviors, this type of fitness functions can help overcoming the bootstrap problem (cf. Sec. 2.1.3). However, it is difficult to define the different stages in such a way that each can build upon the preceding one. Furthermore, it can be hard to decide if one of the temporary behaviors is already evolved properly. As incremental fitness functions involve the hand-coding of intermediate steps of behaviors to be evolved, there is inherently a significant portion of a priori knowledge required. Successful approaches using functional incremental fitness functions can be found in [9, 64, 153].

Tailored fitness functions. These fitness functions contain behavior-based terms as well as aggregate terms, i. e., terms which include observations on a task level, measuring how well a problem has been solved independently of any specific behavioral aspects. For example, when evolving a phototaxis behavior (approaching or avoiding light sources), a possible fitness function might reward controllers that arrive at light sources (or dark places), regardless of their actual actions on a sensoric or actuator level. This would be considered an aggregate term of the fitness function, and the fitness function itself would be considered aggregate if it was the only term. If the calculation additionally involved a behavioral term, such as rewarding robots which face toward (or away from) a light source, the fitness function would be considered tailored. Another example of a tailored fitness function is the evolution of Object Transport where in an aggregate part the fitness function might measure if an object has been transported properly, and additionally reward aspects such as being close to an object or pushing it in the right direction in its behavioral part. Examples from literature include [97, 155, 171]. The "Gate Passing" behavior in this thesis is evolved using a tailored fitness function, cf. Chap. 4.

Environmental incremental fitness functions. Fitness functions of this type are similar to functional incremental fitness functions as they involve increasing the difficulty of gaining fitness during an evolutionary run. However, in contrast to the former, the complexity of the fitness function is not directly increased by adding or complexifying terms, but the environment in which the robots operate is made more complex. For example, after a first solution to a phototaxis task has been found, light sources might begin to move, or more obstacles might be added to make it more difficult to approach the light. In terms of explicity of the selection process (cf. Sec. 2.1.1), this type of fitness functions involves a substantial implicit part which has to be designed carefully to make the evolutionary process successful. Further examples can be found in [9, 128, 135, 149].

Competitive and co-competitive selection. This category refers, rather than just to a type of fitness functions, to a specific type of selection including an according set of fitness functions. Competitive and co-competitive selection exploits direct competition between members of an evolving population or several co-evolving populations. In contrast to other ER scenarios which in almost all cases involve controllers competing with each other in the sense that their measured fitness levels are compared to each other during selection, in competitive evolution the success of one robot directly depends on its capability of reducing the success of another robot. An example is a limited-resources scenario where robots fight for food to survive. In co-competitive selection, the population is divided into several subpopulations which have conflicting target behaviors. Designed properly, this can lead to an "arms race" between the competing subpopulations. A popular and particularly well-studied case of co-competitive selection is given in predator-prey scenarios. More examples of competitive and co-competitive selection can be found in [20, 26, 27, 138, 141].

Aggregate fitness functions. Fitness functions of this type rate behaviors based on the success or failure of a robot with respect to the task to complete, not considering any behavioral aspects of how the task has been completed. For example, a (naïve) aggregate fitness function for Gap Crossing might reward robots according to the time they require to cross a gap. This type of fitness calculation, sometimes called all-in-one evaluation, reduces human bias induced by a priori knowledge to a minimum by aggregating the evaluation of all of the robot's behaviors into a single task-based term. Truly aggregate fitness functions are difficult to design, and such approaches often end up in trivial behavior due to insurmountable success requirements. In the above example of Gap Crossing, a robot population is likely to never achieve the capability of crossing the gap at all which would result in no robot ever gaining superior fitness in comparison to the other robots (cf. bootstrap problem, below). There are few examples of a successful use of aggregate fitness functions in the literature; some can be found in [74, 80, 115].

2.1.3 The Bootstrap Problem

A properly working selection operator relies on the ability of the fitness function to distinguish the qualities of the behaviors in a given population. Otherwise, for example, if all behaviors of a population obtain zero fitness, the evolutionary process is not directed, but purely random. For sophisticated fitness functions, the *bootstrap problem* denotes the unlikeliness of the occurrence of any behavior which achieves a significantly superior fitness compared to the rest of the population when starting in a population of robots with similar behavioral qualities. As in the beginning of a run all robots are expected to have similar fitness, selection cannot direct evolution in that phase, but the first significant improvement has to be achieved by random mutations only (including mutations which result from recombination). The same is true for local optima, i. e., *neutral fitness plateaus* where the whole population has evolved a part of the behavioral requirements given by the fitness function, and, therefore, all robots are indistinguishable by fitness. Hence, when all individuals of a population reside on a neutral fitness plateau, selection is not capable of working properly, leaving mutation as the only operator capable of producing improvements. If an improvement demanded by the fitness function is too hard to be achieved by mutation, the population remains stuck on the plateau. For evolution to work, fitness functions should be designed in a way that a behavior with an above-average rating is likely to occur in any phase of evolution. For example, Urzelai et al. report that a phototaxis behavior in a complex arena can be evolved more easily when it is preceded by the evolution of a CA behavior [195]. However, when evolving complex behaviors it may be very challenging to avoid (premature) situations where the occurrence of a significant improvement is highly improbable.

The bootstrap problem is often considered the main impediment to evolving complex behavior, and consequently it is currently one of the main challenges of ER. A possible solution to the bootstrap problem is to use incremental fitness functions (functional or environmental) which reward intermediate steps of mediocre behavior, until finally leading to the desired target behavior. This approach has been successfully used in several studies [10, 56, 64, 99, 128, 133, 135, 149, 153, 194, 195]. However, incremental approaches require a great amount of a priori knowledge regarding the problem to solve or a complex augmenting environment which both, if inaccurately designed, can lead the evolutionary process to a dead end. Moreover, incremental evolution usually requires to precisely order the different sub-tasks, and to determine when to switch from a sub-task to another. For these reasons, incremental approaches are not easily generalizable to more complex or more open tasks.

Another approach to overcome the bootstrap problem is to efficiently explore the neighborhood of the currently evolved candidate behaviors for one with an above-average fitness. Similar ideas have been widely investigated in EC as *diversity-preserving mechanisms* [55, 169] and *life-time learning* [71, 147]. Typically they rely on a distance measure between genotypes or phenotypes. Recently, another approach has been followed to directly measure behavioral diversity in ER with the purpose of supporting population diversity [134]; this approach relies on measuring distances between behaviors rather than between genotypes or phenotypes. In Chap. 5 of this thesis, novel genotypic encodings and mutation operators are proposed which allow for evolving the GPM together with the robot behavior. As a side effect, this approach leads to a higher population diversity which, in turn, can reduce the bootstrap problem. As discussed in Chap. 5, this is a possible explanation (although not the most satisfactory one) of the improvements observed in comparison to the approach studied in Chap. 4.

2.1.4 The Reality Gap

Another well-known problem in ER is the transfer of simulated results into real-world applications [83, 85]. It has been shown for many pairs of simulation environments and corresponding real robot hardware that controllers which have been evolved in simulation can have a very different behavior when transferred to real robots [142]. This problem, often referred to as the *reality gap*, arises from several real-world factors which are inherently hard to simulate; among these are

- unknown (physical or structural) details of the real environment and the real robots,
- known properties which are infeasible to simulate due to computational costs,
- unpredictable motoric and sensory differences between robots,
- mechanical and software failures.

Furthermore, small variances between simulation and reality can quickly lead to great differences in terms of robot positions and angles. Particularly, this makes behaviors virtually unsimulatable which are relying on exact distance and turning instructions. Let, for example, a robot be instructed to turn for 45 degrees and drive straight ahead thenceforward. Obviously, it can soon reach significantly different positions if the actual turn which is accomplished by its motors is in a range between 40 and 50 degrees (where 5 degrees of variance due to mechanical properties are a rather moderate estimation when working with current mobile robots).

On the other hand, there are obvious advantages in terms of computational and cost efficiency of utilizing simulation when evolving robot controllers. Therefore, there have been different attempts to avoid the reality gap. A common method is to use a sophisticated simulation engine with physics computation and artificially added noise to create unpredicted situations. To deal with the high computational complexity of these simulations, an established method is to use different levels of detail depending on the purpose of the simulation [2, 3, 83]. Another way is to combine evolution in simulation with evolution on real robots [199]. Following this approach, it is possible to use the measured differences between simulation and reality as a feedback in an "anticipation" mechanism to adjust the simulation engine [63]. Another idea is to adjust evolutionary parameters and operators in a rather simple simulation and to actually evolve behaviors onboard of real robots. This approach avoids the reality gap under the reasonable assumption that, although motoric and sensory details might be hard to simulate, the mechanisms of evolution work on real robots basically in the same way as in simulation [199]. The experiments described in this thesis are based on this assumption; Chap. 4 provides further evidence for its validity by transferring simulation results to real robots. However, this approach requires to actually evolve behaviors in an expensive real-world scenario after having tested the basic process in simulation. Chap. 6 proposes and discusses a method for predicting the success of an evolutionary run in a real-world environment based on simulation data.

2.1.5 Decentralized Online Evolution in ESR

Many of the concepts presented in this thesis are in principle applicable to various evolutionary approaches within and outside of ER. For example, the FSM as a formal model has been used to evolve programs for several interesting tasks, such as the prediction of prime numbers, in the field of Evolutionary Programming [50]. Here, however, all concepts are applied to a robot swarm in an embedded evolutionary setting with the purpose to learn behaviors in an online and onboard manner.

As outlined in Sec. 1.2, most approaches in ER and ESR can be categorized into offline or online techniques. In an offline setting a robotic system (i. e., swarm or single robot) learns to solve a desired task in an artificial environment (real or simulated). Subsequently, the evolved behavior is hard-coded into the controllers to solve the task in a real environment. In online settings a robotic system has to learn in an environment which it simultaneously has to solve the task in. This means that during a run, currently evolved behavior is evaluated by observing its performance on the task to solve. In contrast to offline evolution, it is usually not possible (or infeasible) to wait for an eventually evolved behavior of sufficiently high quality, but intermediate behaviors have to be employed and evaluated, too. The requirement of learning behaviors online is given, for example, when robots have to adapt quickly to new and possibly unknown situations or when they have to learn how to deal with novel objectives which occur randomly during a run. Particularly, when robots explore new areas which are at least partially unknown beforehand, objectives may change constantly.

Typically to ESR, a further distinction is made between approaches which utilize observing computers (centralized) and those which do not (decentralized or onboard). In a decentralized system, every robot has to make all decisions based on local observations, only. In most examples from literature, centralized techniques are employed in combination with offline evolution while online evolution accompanies decentralized techniques. Popular examples of centralized offline evolution include, e. g., [1, 4, 53, 59, 60, 61, 178, 179, 192, 198]; popular examples of decentralized online evolution include, e. g., [19, 68, 70, 93, 94, 130, 170, 184, 185, 186, 187]. There exist some examples which are best categorized as decentralized offline evolution [69, 182], but, to the knowledge of the author, no work exists so far combining centralized with online evolution.

Evolutionary approaches that are on the decentralized and online side of the classification scheme (such as those proposed in this thesis) are sometimes called *Embodied Evolution (EE)* [200]. One of the first works in EE has been the "Probabilistic Gene Transfer Algorithm" [200] as well as the work of Simões and Barone [36]. Both approaches use an ANN model as a control mechanism to evolve robotic behavior (cf. discussion in Sec. 2.1.6). Embodied evolution can be implemented in simulation as well as on real robot platforms, and it can be applied to a variety of different scenarios. Due to the decentralization, embodied approaches scale well to large swarms of robots, and due to the online property, the main focus is on finally achieving an adaptation to a real environment, rather than to a simulated environment or an environment in a laboratory experiment. Therefore, simulation and laboratory experiments are used in EE to study evolutionary mechanisms only, but not to evolve behaviors for the purpose of using them as static controllers in real-world environments.

Offline and centralized techniques are known to converge more quickly to good solutions for several problem classes than EE techniques do, but there is a broad range of applications for which they cannot be used. Particularly in swarm robotics, there are innumerable important applications for which EE appears to be a better choice. Whenever swarms of robots are supposed to solve tasks in new and (partially) unknown areas which are not accessible to humans or other global observers, a decentralized approach is required; and whenever such an area contains unknown sub-challenges interfering with the global goal, or when unforeseeable changes to the environment are likely to occur during the robots' mission, an online approach is necessary.

Real-world applications range from recovering victims in disaster areas or unmanned discovery of the surface of the sea or of other planets, to healing diseases by injecting a swarm of nano-robots into the human body. While such applications seem rather futuristic today, their theoretical feasibility in terms of available robot hardware is steadily increasing. Therefore, ER research has to keep up to prospectively be able to provide the required software for the control of future swarms of robots (or to play an appropriate part in a more general control framework – which is more likely, from today's perspective).

Beyond the rather uncritical examples listed above (in terms of their intended purposes), there are many imaginable applications which might harm people or the earth's environment in various ways. Applications related to weapons, military equipment, spy out of people, which is even today technologically feasible in several ways, and many others are not discussed here, although the proposed techniques cannot be prevented from being used in these areas. However, profound ethical considerations are beyond the scope of this thesis and the author's competence. Nevertheless, thorough analyses of the ethical implications of all planned applications are inevitable and should be ubiquitous for all technologies that have a potential to majorly influence the wellbeing of people. Particularly, mere market analyses cannot be an appropriate replacement for such analyses.

2.1.6 Evolvability, Controller Representation and the Genotype-Phenotype Mapping

According to current biological understanding, in natural evolution there is a rather strict separation between the *genotypic* world where mutation and recombination take place and many characteristics of an organism are encoded, and the *phenotypic* world where individuals inhabit a niche in the environment and are subject to selection as part of the "survival of the fittest" (or the "struggle for life", as Charles Darwin framed it in the "Origin of Species"). In artificial evolution, this separation can be omitted for simplicity reasons by evolving solely in the solution space or *phenotypic space*. In the case of ER this means that mutation and recombination are directly applied to behavioral programs of robots. Most approaches in ER so far have been performed in this manner due to the limited resources on real robot hardware. However, a well-established approach from EC is to mimic natural evolution by introducing a genotypic representation of an individual, different from the phenotypic representation. As in nature, each genotype then corresponds to a phenotype, i. e., a robot controller, which has to be computed from the genotype by a translation procedure called GPM. Thus, mutation and recombination operations are performed in the *genotypic space* while evaluation and selection take place in the phenotypic space. This differentiation between genotypes and phenotypes has proven to enhance the performance of evolution for several reasons. Due to the possibility of introducing genotypic representations with less syntactical constraints than the corresponding phenotypes have, a "smoother" search space can be accomplished. For example, in a work by Keller and Banzhaf, bit strings are used to genetically encode C programs, using a repair mechanism as part of the GPM to make every string valid [92]. The authors report a significant improvement in terms of quality of the found solutions compared to an approach with phenotypes only, explaining the improvement by the hard syntactical constraints on mutation and recombination when working solely in the phenotypic space. Similarly, the phenotypic space used in this thesis consists of controllers encoded as FSMs (cf. Sec. 2.1.7) which are accompanied by hard syntactical constraints. The according genotypic space introduced in Chap. 5 consists of sequences of integers which are translated into FSMs by using a script including a repair mechanism. Experiments with this static genotypic representation show an improvement in terms of mean fitness and complexity of evolved behaviors, confirming the results of Keller and Banzhaf.

Beyond that, important properties of natural genotypes can be obtained using a GPM which include *expressive power*, *compactness* and *evolvability* [142]. There, ex-

pressive power refers to the capability of a genotypic encoding to capture many different phenotypic characteristics and allow for a variety of potential solutions to be considered by evolution; a compact genotypic encoding is given if the genotypes' length
sub-linearly reflects the complexity of the phenotype; and evolvability is described
in [142] as the capability of an evolutionary system to continually produce improvements by repeatedly exerting the evolutionary operators (note, however, the difference
in the definition of evolvability introduced in Chap. 5).

Usually, the GPM is a non-bijective mapping, meaning that many different genotypes can be translated into the same phenotype. Furthermore, the GPM is not necessarily static, but it can depend on time and the state of the environment. There are
approaches that let the genotype pass a process of development (sometimes called
"meiosis"), which yields more and more complex structures by considering the current states of the organism in development and the environment [67]. Another nature-
mimicking idea is to make the GPM *evolvable*, i.e., itself subject to evolution [142].
Inspired by the natural process of gene expression, where the rules for translating
base sequences into proteins themselves are encoded in the DNA [52], the GPM can
be encoded in a part of the genotype making it implicitly subject to adaptation during
evolution. This can result in an adaptation of the GPM to the fitness landscape given
by the task to evolve and the environmental properties. As a consequence, the indirect
effects of mutation and recombination on the phenotypic level can be adapted to the
fitness landscape, actively directing the search process towards better phenotypes by
skipping areas of low fitness. A practical benefit of this approach is that the mutation
and recombination operators do not have to be adjusted manually before every evolutionary run, but the adaptation is performed automatically, according to the behavior
to evolve and the environment it is evolved in. A related approach is followed in the
well-known *Evolution Strategies* [13] where the mutation strength is encoded as part
of the genotype, thus being evolvable during evolution.

A more general benefit intended to be achieved by the usage of evolvable GPMs is
an increase in evolvability (in the above sense). While in nature the GPM is known to
be a flexible part of evolution, to the effect that the rules determining the translation of
the DNA into proteins are themselves encoded in the DNA, the effects of this flexibility
on evolvability are still discussed [39, 114, 152]. Likewise, it is still an open question if
an evolvable GPM can increase evolvability in ER. This thesis provides evidence that
an evolvable GPM can significantly increase the expected evolutionary success in ER
in terms of behaviors evolved, and that an evolvable GPM can adapt to a given search
space structure as proposed above (cf. Chap. 5). Both these results cannot prove an
increase in evolvability as they can be explained by different effects, too, but they may
serve as building blocks toward a general understanding of evolutionary processes in
ER.

An evolvable GPM allows for evolution to learn structural properties of the search
space, which depend on the behavior to be evolved, and to use this knowledge for
improving evolvability during a run. Fig. 2.3 shows an example of a desirable adapta-

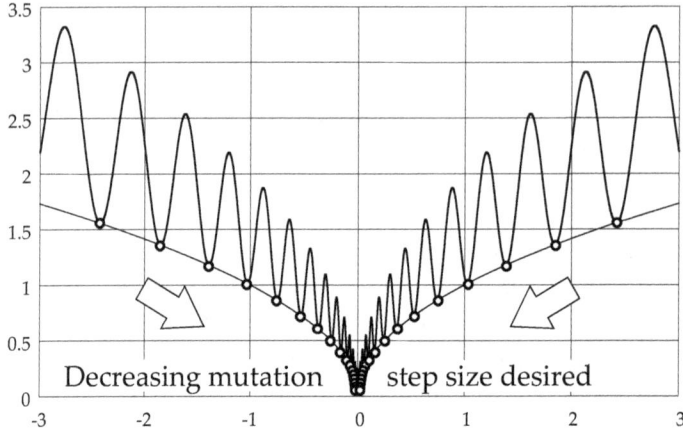

Fig. 2.3. Adaptation of mutation step size in a one-dimensional real-valued search space. The depicted Schaffer 2 function is given by $f : \mathbb{R} \rightarrow \mathbb{R} : x \mapsto |x|^{0.5} \left(\sin^2 \left(50 \cdot |x|^{0.2} \right) + 1 \right)$. When approaching the minimum at $x = 0$, an individual's mutation step size should decrease to skip the local maxima and still approach the global minimum rather than oscillating around it.

tion of the mutation step size when searching the minimum of a "Schaffer 2" function (a typical benchmark function in EC) in a one-dimensional real-valued search space. Due to the flexibility of the GPM, operators can be adapted indirectly by changing the meaning of a part of the genotype, thus changing the impact of the operator when working on that part. In the example, mutation can change during evolution to skip the gaps in the fitness landscape and move on to better fitness regions.

2.1.7 Controller Representation

The term *Controller Representation* (or simply *Controller*) refers to the structure of programs running onboard of robots with the purpose of controlling their behavior. In the past, ANNs have been used most frequently to encode robotic control systems. According to Nelson et al. in 2009, about 40 % of the published results in ER have been achieved using ANNs, while another 30 % used evolvable programming structures, i. e., Genetic Programming [139]. Of the approximately 50 papers reviewed by Nelson et al., only two use FSM-based encodings. The current preference of ANN-based controllers can be attributed to the rather straight-forward implementation of ANNs including a set of well-studied learning operations other than evolution, and simply to the broad success of ANNs in learning various behaviors. However, in most cases these benefits come at the cost of resulting controllers which have a complicated structure and are hard to interpret. In the case of ER, this is due to the randomized changes to the genotype introduced by evolutionary operators such as mutation or recombination which optimize the phenotypic performance without respecting any structural benefits of the genotype. The consequence is that evolved controllers are hard to analyze from a genotypic point of view. While it may seem plausible by observation that a robot has learned a certain behavior, is is usually not possible to prove it. This comes as an additional difficulty to the problem observed by Braitenberg that it is hard to predict robotic behavior solely from knowledge of the controller (cf. Chap. 1, [17]). While this may seem acceptable in some areas, there are many robotic applications for which it is crucial to guarantee that tasks are solved every time with the same precision or in the same manner (e. g., in medical applications where errors can cost lives or on space missions where erroneous behavior can be very expensive).

On the other hand, there are several successful approaches in EC which rely on the rather simple theory of FSMs, e. g., [48, 49, 50, 51, 177]. In ER, there are also a few successful approaches reported using FSMs, e. g., [91, 208]. In [91], the authors call their method Genetic Programming due to the genetic operations performed on the controllers. However, the controllers are built from simple if-then-else blocks triggering atomic actions based on sensoric information, which in essence can be represented by an FSM. (Considering similar cases, there might be a higher number of FSM-based approaches than assumed, as some authors might call their methods differently.) In both publications, the authors report that controllers were generated in simulation and then successfully tested on real robots and in simulation.

The semantics of FSMs is much more comprehensive than the semantics of ANNs leading to controllers that can be analyzed automatically for various non-trivial properties. This is the main reason why in this thesis FSMs have been deployed as controllers, based on a newly proposed controller model called *Moore Automaton for Robot Behavior (MARB)* [102, 105, 106]. Beyond this thesis, the MARB model has proven successful in a medical application (in a non-ER setting) studied by Jung et al. who have used MARBs to generate controllers for a robotic exoskeleton walking assistant with crutches for paraplegic patients [88].

The price for the analyzability of FSMs is their lower expressive power and, when used in a straight forward way, their lower evolvability compared to ANNs which are both Turing-complete and highly evolvable in many scenarios. However, as reported many times, for example by Braitenberg, complex control structures are not necessarily required to encode complex behaviors. Particularly, the complex behaviors which Braitenberg established by direct connections from sensors to motors (cf. Chap. 1), are performed by even simpler control structures than FSMs. Furthermore, by far the most of all existing controllers learned with ANNs could as well be represented as FSMs meaning that the expressive power of ANNs is generally not exploited in today's robotic controllers. The second drawback, namely the lower evolvability of FSMs, is a more serious problem, and searching for a solution to it is a main topic of this thesis. It is discussed in Chap. 4 where an approach without a distinction between genotypic and phenotypic representation is presented, and in Chap. 5 where a genotypic representation for FSM-based controllers and a completely evolvable GPM are introduced. In the approach without a genotypic representation, the structural benefits of the MARB model are exploited by a mutation operator which is designed to "harden" those parts of an FSM which are expected to be involved in good behavior. Other parts of the FSM stay loosely connected and can get deleted within few mutations. In this way, the complexity of the automaton gets adapted to the complexity of the task to be learned (similarly to an ANN-based approach by Stanley and Miikkulainen [180]). In the approach with an evolvable GPM, the translation from genotypes to phenotypes is supposed to evolve along with the behavior making mutations and recombinations adaptable to the search space structure, as described above.

2.1.8 Recombination Operators

Recombination is rather rarely used in ER as it tends to hardly improve evolution by being laborious to design and set up [142]. Nevertheless, there are studies that show the successful adoption of recombination operators in ER. For example, Spears et al. evolve FSMs for the solution of resource protection problems by representing a geno-type as a matrix assigning to each state/input pair an action and a next state (essentially a Mealy machine) [177]. They observe that evolution is capable of finding the appropriate number of states necessary to learn a behavior of a certain complexity. Furthermore, they report that a recombination operator has a great positive influence on evolution. In Chap. 4, a similar recombination operator is used on controllers represented as MARBs without involving an explicit GPM. The main result, however, is that it has a rather small effect (if any) on evolution. Moreover, the effect is positive in certain specific parameter combinations only, by even being harmful in others.

2.1.9 Success Prediction in ESR

A key problem in ESR is to accurately select controllers for offspring production with respect to performing a desired behavior [142]. This means that "better" controllers in terms of the desired behavioral qualities should have a higher chance of being selected than "worse" ones. However, it is usually not possible to grade arbitrary evolved controllers detached from the environment in which the desired task has to be accomplished. Therefore, all ESR scenarios require the existence of an environment (real, simulated or even further abstracted) where the controllers can be tested in. Using an environment to establish the quality of controllers (i. e., their explicit and implicit fitness) makes ESR more closely related to natural evolution than most classic EC approaches (cf. the two different views on ESR in Sec. 2.1.1).

 Requiring a complex environment for fitness evaluation makes it hard to predict the success probability of a given ESR scenario, i. e., the probability that a behavior capable to solve a given task will be evolved in an acceptable time. On the other hand, such a prediction is essential if failures can yield to expensive hardware damage or even harm humans. Due to the reality gap, this is a common case for many desired applications. The need for evolving online or at least in an environment designed as similar to the real environment as possible tends to make experiments expensive.

There has been a lot of theoretical work in the fields of classic EC, spatially structured evolutionary algorithms, evolutionary optimization in uncertain or noisy environments, and ER aiming at the calculation of the expected success of an evolutionary run, and, as a result, on the calculation of the required settings to increase the chance of success [5, 62, 77, 86, 95, 151, 154, 166]. Furthermore, there are well-established models of natural processes from the field of Evolutionary Biology [52], and interdisciplinary concepts such as Genetic Drift [98] and Schema Theory [77]. However, there is, to the knowledge of the author, so far no theoretical model which includes implicit (environmental) selection as well as explicit (fitness based) selection in the calculation of success. The combination of these two types of selection is considered typical to ESR [142], and occurs neither in classic EC nor in Evolutionary Biology separately.

In Chap. 6, the influence of environments on the evolution process in ESR is studied theoretically and experimentally. A hybrid mathematical and data-driven model based on Markov chains is presented that can be used to predict the success of an ESR run. By successively adding more and more data from simulation or laboratory experiments, the prediction can be improved to a desired level of accuracy before performing an expensive real-world run. The model considers implicit selection properties and the *selection confidence* of a system, i. e., a measure of the probability of selecting the "better" out of two different robots in terms of their desired behavioral properties. The latter is related, but not identical, to explicit fitness. The model calculates the expected success rate of a given ESR scenario based on a series of "fingerprints", i. e., abstracted selection data typical to a scenario derived from available simulation or real-robot experiments. So far, experiments support the hypothesis that the model can provide accurate success estimations, although more experiments are required to confirm this statement, cf. Sec. 6.4. However, it is important to note that the model as presented in this thesis is still at a basic stage and further work is necessary to make it applicable to real-world problems, cf. discussion in Chap. 6.

2.2 Agent-based Simulation

The field of ABS has experienced a quick growth since the early 1990s. It has been established in many fields of academic research as an indispensable technology to yield insights that neither formal reasoning nor empirical testing are able to achieve [173]. ABS offers intriguing ways to model superordinate effects which emerge from local interactions of individual agent behaviors. Applications range broadly from economic examinations of stock trading policies through traffic planning and social interaction modeling to the development of complex behavior which includes robot evolution in

ESR. In the last two decades, various ABS platforms have been developed, some of which have specific properties intended to suit a particular problem area while others serve as general-purpose frameworks. The former include MAML for social sciences, NetLogo, StarLogo and, more recently, ReLogo for education, Jade, NeSSi and OMNET++ for networking, Breve and Repast for artificial intelligence, MatLab, and OpenStarLogo for natural sciences, and Aimsun and OBEUS for urban/traffic simulation. Among the latter, i. e., the general-purpose ABS frameworks, there are programs of different powerfulness and complexity, for example, Anylogic, Ascape, DeX, EcoLab, Madkit, Magsy, Mason, NetLogo, ReLogo, Repast, Swarm, SOAR and many more. These lists intersect, as some simulations have originally been developed for a specific application area and generalized later, and they are by far not limited to the given examples. These simulations mostly run on popular operating systems such as Windows, Linux or Mac OS including a Java Runtime Environment or GNU C/C++ Compiler. Many of these frameworks are publicly available for free, at least for non-commercial usage [140, 157, 172].

Depending on the focus of the users, each of these simulation frameworks has its advantages and drawbacks. Some fit in very special application areas only, such as the microscopic traffic modeling framework Aimsun [8, 22], while others are flexibly adjustable to various types of applications, such as Netlogo [174], Mason [117], Swarm [31, 129], and Repast [29, 144], as well as the merger of Repast and NetLogo, ReLogo [119]. In addition to those, many others have emerged in the last years driven by specific requirements and emphases of the developers. Particularly, great progress has been made in developing more and more practical ABS toolkits. These toolkits enable individual developers to implement considerably more powerful and complex applications using the agent paradigm than before. Simultaneously, there has been a quick increase in the number of such toolkits [140]. There have been efforts in organizing the available ABS frameworks by comparing several of their functional and non-functional properties with each other. However, such efforts, due to the large number of available simulations, either focus on few programs or give rough overviews only. In [23], the eight toolkits Swarm, Mason, Repast, StarLogo, NetLogo, Obeus, AgentSheets, and AnyLogic are described and compared to each other; MASON, Repast, NetLogo and SWARM are analyzed, and compared to each other in [157]; Tobias and Hofmann examine the four toolkits Repast, Swarm, Quicksilver, and VSEit [189]; Serenko et al. examine 20 toolkits with a focus on instructors' satisfaction when using the software in academic courses [172]; Nikolai and Madey perform a broad examination and categorization of a large list of ABS toolkits [140].

However, it has been pointed out recently that ABS is still at an "infancy stage" and more work in analytical development, optimization, and validation is needed [24]. Other simulation paradigms, such as discrete-event simulation, system dynamics, or finite-element-based simulation are claimed to be different in this respect as they have already developed a rich set of theories and practices. One issue that is particularly important for big research projects, is the development of structured and reusable code.

Structured code can be the foundation of an error-free implementation, especially if unskilled programmers are involved. This is, for example, frequently the case in academic research where students are involved within the scope of a thesis or a laboratory course, but also in many other projects which include people without a software engineering background. While there exist popular easy-to-learn ABS toolkits such as MASON or NetLogo, their architectures do not guide unskilled users actively to well-structured code, but leave it to their own competency. This can lead to code that is hard to reuse, and understand, and, in turn, to potentially erroneous implementations. It is still challenging for non-experts to generate suitable simulation models and implement them in a structured and reusable way using most of the current simulation frameworks.

Furthermore, considering the multitude of available simulation programs, it is challenging to select for a given task an appropriate ABS framework to work with. Beyond task-inherent constraints, the existence of convenient features has to be taken into account which range from practical functionalities, such as automatic chart generation or an intuitive user interface, to flexibility, extendability and reusability [24, 120]. Modularity is another key feature, as different programmers should be able to work on one large project without interfering with each other while still being able to reuse each others' code. Moreover, it should be possible to reconfigure the programmed modules from use case to use case and to reuse modules in simulations different from the one they were originally designed for. Another important requirement is efficiency which concerns time consumption as well as memory usage because long or complex simulations tend to occupy large amounts of memory. A support for automatically outsourcing simulation runs, for example to a cloud service, is therefore desirable, too. Overall, from a non-expert perspective, the seeming benefit of a huge variety of available ABS frameworks turns out to be overwhelming, and the difficulty in generating structured code is a major reason for the break down of software projects involving unskilled users (at least according to the author's experience which is limited to the area of academic research).

In this thesis a novel architecture for programming agent simulations is introduced, which is called SPI. The SPI is particularly suitable to guide student programmers with little experience to well-structured and reusable simulation components. It introduces an intermediate layer, the *Scheduler/Plugin Layer (SPL)*, between the simulation engine and the simulation model which contains all types of functionality required for a simulation, but logically separable from the simulation model. This includes (but is by far not limited to) visualization, probes, statistics calculations, logging, scheduling, API to other programming languages etc. The SPL offers a uniform way of extracting functionalities from the simulation model which logically do not belong there. Due to the rigorous detachment of the simulation model from the simulation engine, which can be bridged solely by using the SPL, users are forced to implement not only on the model side, but also modules on the SPL side. The extracted modules can be attached and detached flexibly, and used in completely different scenarios, making the code reusable. The modules that can be placed in the SPL are called *plugins* in general, and *schedulers* if they are mainly used for scheduling purposes. The SPI architecture can be implemented in a simple way as an extension to most state-of-the-art simulation frameworks. It has been inspired by the *Model/View/Controller (MVC)* architectural pattern [21] and can be seen as a generalization of that pattern.

For this thesis, the SPI has been implemented as part of the ABS framework EAS [109]. It has been created out of the requirement of a general-purpose simulation framework that could be used in projects where academic research is combined with student work and that supports all the above-mentioned criteria in an active and intuitive way with a focus on modularity and reusability. This means that the desired framework should not only allow implementations to have these properties (which is the case for many of the currently available simulation frameworks), but the programming architecture should guide particularly unskilled users in an intuitive way to a well-structured implementation.

EAS has a slim simulation core which, due to its size and simplicity can be considered virtually error-free. It provides a notion of time and the SPL interface that establishes a connection to simulation models. All additional functionalities are implemented as part of a simulation model or a plugin/scheduler. By removing all unnecessary functionalities from the simulation core, the time and space efficiency is left to the end-user programmers meaning that the simulation core does not constitute a bottleneck of speed or memory.

Another feature of EAS is the direct connection to the job scheduling system JoSchKa [15]. JoSchKa can be used to schedule a magnitude of jobs to work on several local computers in parallel or it can establish a connection to a cloud service to use even greater computing resources over the internet. Using JoSchKa, EAS can run different simulation runs on many computers in parallel or computationally expensive simulations on a single powerful remote computer.

EAS is programmed in Java and can be used as a general-purpose simulation framework. It is a rather new simulation platform being developed in its latest stage for about three years (in 2014). Nevertheless, many common simulation scenarios (including 2D grid and continuous simulations as well as 2D and 3D physics simulations) and a set of important plugins (such as visualization, chart generation, trajectory drawing, logging etc.) are already implemented in EAS. It is an open-source project running under a creative commons license, and it can be downloaded including all sources and documentation from sourceforge[2].

In the next chapter, a detailed description of the SPI and the EAS Framework is given, and the results of a study are presented which compares EAS with the state-of-the-art ABS toolkits MASON and NetLogo.

---

2 EAS on sourceforge: http://sourceforge.net/projects/eas-framework [101]

3 The Easy Agent Simulation

Preamble: *There exists a huge amount of groundwork in the field of agent-based simulation software. Cynthia Nikolai and Gregory Madey count in their 2009 survey paper as many as 53 agent-based modeling platforms [140], a list which still contains general-purpose frameworks only. There are simulations for a variety of application areas (including scientific applications), written in different programming languages, protected under various license types, and many being free for non-commercial use. Then why does this thesis add another simulation to the list, instead of using an existing one? The answer is twofold: Firstly, it is an enriching and enjoyable experience to implement a simulation platform from scratch. Secondly, and more scientifically, by focussing on the specific requirements of a doctoral thesis such a trial can lead to new concepts which have benefits over the existing frameworks. In this case, a new concept has been discovered that helps especially non-expert users to quickly learn implementing well-structured and reusable simulation code.*

Major growth in the field of ABS since the early 1990s has led to a paradigm shift in academic research in various domains and it accompanied establishing simulation as a third pillar of science beside theory and experiment [173]. As a consequence, simulation frameworks are used with increasing frequency by a heterogeneous group of people in terms of programming skills, ranging from high-skilled software engineers to non-specialist students. Frequently, both these groups work together on the same project which raises the requirement for simulations which allow for complex scenarios to be simulated by still yielding well-structured and reusable code, even when produced by unskilled programmers.

This chapter presents a general architectural pattern (first proposed in [109]) as well as a new Java-based simulation software implementing this pattern, both aiming at providing a powerful framework which guides unskilled users quickly and safely to implementing well-structured simulations. The next section briefly describes the historical background of the simulation software (called EAS Framework). Sec. 3.2 describes the basic architectural concept, and Sec. 3.3 gives some details about the actual implementation of the EAS Framework in Java. Sec. 3.4 describes a study comparing the approach to two state-of-the-art simulation programs. Sec. 3.5 concludes the chapter and gives some hints to implementing the architectural concept within existing simulation programs other than the EAS Framework.

Fig. 3.1. Photography of a swarm of Jasmine IIIp robots. The Jasmine IIIp robot has been simulated using the EAS Framework to perform various kinds of evolutionary and other experiments as described in the subsequent chapters. The robot is sized about $26 \times 26 \times 26 \, mm^3$; a 1-Euro coin is placed next to a robot to illustrate its size.

3.1 History of the Easy Agent Simulation Framework

The first basic version of the EAS Framework has been developed for a diploma thesis in 2007 [100] within the scope of the proposal process for the EU projects "Symbrion" and "Replicator" [93] at the University of Stuttgart. At first, it had the sole specific purpose of simulating swarms of *Jasmine IIIp* robots, cf. Fig. 3.1, with programs encoded as FSMs (cf. *Moore Automaton for Robot Behavior*, Chap. 4). The program has, thus, been called *Finite Moore Generator (FMG)*. It is described in detail in the diploma thesis and a 2008 research paper [105], including several experiments with real Jasmine IIIp robots.

At the beginning of the work on this doctoral thesis, FMG has been extended to allow evolution experiments to be performed purely in simulation (as real robots were not available then). This led to a first publication on evolution using FSMs in simulation [102]. Subsequently, FMG has been further extended and used for various experiments in artificial evolution and swarm robotics, cf. Chapters 4, 5 and 6 (to a large extent, these experiments have been performed with an FMG version embedded in the

Fig. 3.2. Two photographies of several Wanda robots. The Wanda robot has been simulated with the EAS Framework to perform mostly evolutionary experiments with controllers based on FSMs. In Chap. 4, it is used in the real-robot experiments, cf. Sec. 4.4. The robot has a diameter of about 40 mm and a height of about 60 mm.

more general simulation program which eventually became the current EAS Framework, cf. below).

In 2009, the robot platform *Wanda* has been developed in cooperation with the IPR at the KIT [96], cf. Fig. 3.2. As Wanda robots became available for experiments, FMG has been adapted to simulate these robots, too. At this point, FMG had become a large and complex program, complicated to use by anyone other than its author. At the same time, an advanced laboratory course on the topic of evolutionary swarm robotics, called *Organic Computing: Learning Robots (OCLR)*, has been initiated. There, students were supposed to implement robot behavior in simulation as well as on real robots. For this purpose, a basic version of the SPI has been developed and implemented within FMG with the purpose of allowing beginners to quickly start programming individual simulation scenarios by simultaneously avoid altering any parts of the simulation core. After having tested the concept during two semesters of the laboratory course and in the scope of several student theses, another major rewrite of FMG has been performed leading to the final version of the SPI described here. At this time it has become possible to use the simulation in many different do-

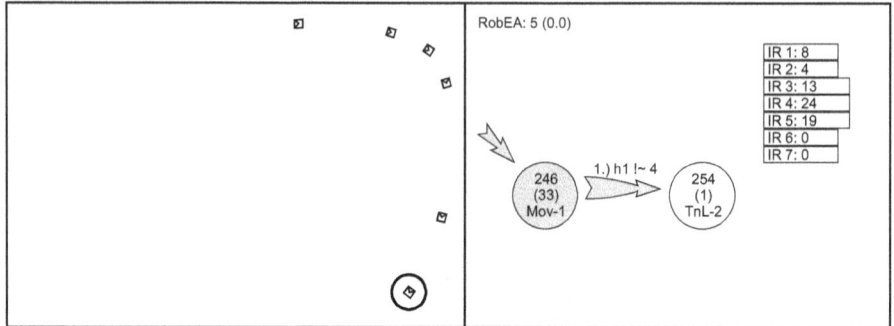

Fig. 3.3. Live simulation view of an FMG run embedded in the EAS Framework. Several Jasmine IIIp robots are shown on the left side, visualized by a fundamental plugin of the EAS Framework called "videoplugin". The MARB controller of the selected robot (marked by a circle) is implemented as a generic agent brain and visualized on the right side. The selected robot's sensory perception (in terms of numeric values of its seven infra red sensors, cf. Chap. 4) is depicted, too, on the right side.

mains for various scenarios reaching from abstract agent simulations to simulations in 2D and 3D with or without physics simulation. The classic FMG simulation has been embedded as one specific scenario in the new simulation program, cf. Fig. 3.3. Due to its purpose of being easily understandable for beginners and still sufficiently flexible to allow implementing various types of simulation scenarios, the name "Easy Agent Simulation" has been given to the overall program.

EAS has been studied during one semester of OCLR in a real student teaching situation with respect to its ability to lead unskilled users to well-structured simulation code. The code implemented in EAS has been related to code implemented by the same students using the state-of-the-art ABS toolkits NetLogo and MASON. The results are presented below in this chapter, and in a paper published at the *Winter Simulation Conference* in 2012 [109]. There, suggestions are given of how the SPI can be implemented in other existing ABS toolkits making the use of the EAS Framework optional.

Furthermore, all ESR experiments in this thesis have been performed using either FMG in the early stages or the FMG implementation embedded within the EAS Framework. Therefore, the simulation platform has been uniform for all experiments. As the exact implementation of a simulation can highly influence the outcomes of experiments, particularly if randomized processes are involved, it has been an important constraint to provide a complete and precise backwards compatibility when adding new implementations. Randomly chosen test sets indicate that experiments performed using early versions of the implementation are exactly reproducible using current versions. However, as this claim is not proven, every experimental setting in this thesis has been described on an abstract level avoiding implementation details.

Particularly, statistically significant characteristics of the results can be understood as completely independent of the underlying simulation program.

3.2 Basic Idea and Architectural Concept

The EAS Framework is implemented in Java and serves as a general-purpose simulation program. EAS as well as the SPI concept are, in principle, not only applicable to ABS, but can, for example, also be used for synchronous or asynchronous discrete-event simulations. However, as the name indicates, a special emphasis has been put on ABS, and practically all implementations available so far are agent-based. Therefore, while the EAS architecture and its implementation in Java are described at a universal level, all examples in this thesis as well as the implementations performed in the scope of the comparative study are under the agent paradigm.

3.2.1 Overview

As described above, EAS has been developed for research as well as teaching purposes. The SPI architecture and its implementation in the EAS Framework arose from the need in academic research for a simulation program which simplifies the fusion of simulation results originating from postgraduate research studies on the one hand, and students' theses or lab courses on the other hand. Using a novel programming architecture, the SPI allows for a simple and flexible agent and environment creation, and offers the ability of adding external functionality by a generic interface called *plugin*. At the same time, the architecture has been designed to be as lightweight as possible allowing to write memory and time efficient simulations which may involve real-time demands. Currently available implementations range from abstract agent simulations without any connection to a physical environment over grid or continuous space simulations with pseudo-physical agent-environment interactions to simulations with complex physics in a two or three dimensional space. Furthermore, EAS has the ability of automatically creating job packages for the job scheduling system *JoSchKa* [16] which, among other features, allows to easily connect to cloud services.

3.2.2 Preliminaries

This section introduces basic terms used throughout this chapter. Some of the terms are adjusted to the proposed SPI architecture and may be defined differently than in the literature. For example, a distinction is made between a "time instant", a "tick" and an "event" as these notions are explicitly distinguished within the EAS Framework. Technically, any of these terms can be considered an "event" which is frequently done in the literature.

Definition 3.1 (*Basic terms*)

A **notification** is an object that is passed from one part of a simulation program to another providing information about state changes. Notifications are used to propagate the progression of time and the occurrence of events.

A **time instant** is a point in continuous time space $\mathbb{R}$, represented by a positive real value.

A **tick** is a discrete time instant in $\mathbb{N}_0 \subset \mathbb{R}$, represented by a positive integer value.

An **event** is an object of a specific type (class EASEvent in EAS) that can be broadcast at any time during the runtime of a simulation. Events may be timeless or bound to a time instant, and they cover, in that respect, a superclass of the information given by time instants (however, in EAS there is no inheritance relation implemented for flexibility and efficiency reasons).

The term **simulation core** denotes the basic skeleton of the simulation program, minimally required to run a simulation (it contains several classes/packages in EAS).

The **simulation engine** is that part of a simulation core which provides a notion of time and creates notifications of events and the progression of time (class SimulationTime in EAS).

The term **simulation model** denotes the implementation of the world that is simulated including rules of state changes due to progression in time and events. A simulation model typically includes an *environment* and several *agents* acting in it. (In EAS, the simulation model contains several classes/packages. Every specific simulation run involves one distinguished entrance point to the simulation model, an object of type EASRunnable – typically representing the environment. It is referred to as **main runnable** in the following, and it receives notifications from the simulation engine (more precisely, over the detour of a scheduler, see below) to pass them to other classes of the simulation model – say agents, if required).

The term **scheduling** denotes, in its most general form, the potentially manifold assignment of objects from some arbitrary finite set $S = \{O_1, \ldots, O_n\}$ to points in time, i.e., time instants; it can be expressed as a mapping $a : S \to \mathbb{R}$. In the context of EAS, S is the set of invokable methods within the simulation model. A **scheduler** is a program or program part that chooses a mapping a from the set of all schedulings $Sched = \{a' \mid a'$ is a mapping $S \to \mathbb{R}\}$ and performs the according method invocations at the appropriate time instants based on notifications from the simulation engine. A simulation program is called **completely schedulable** if it allows for every scheduling $a \in Sched$ the implementation of a scheduler that induces the mapping a.

3.2.3 Classification of the Architecture

The basic idea underlying the SPI architecture is to introduce an intermediate layer, the SPL – hosting schedulers and plugins, between the simulation engine and the simulation model. The SPL can be used to aggregate all types of functionalities in a standardized way which influence or observe a simulation run and belong neither to the simulation engine, nor, by not being part of the simulated world, to the simulation model. The consideration to use an SPL as part of a simulation program originates from the following observations which are valid for a variety of current state-of-the-art ABS frameworks:

1. There are several types of typical functionalities that do not belong to the simulation engine, but should as well be separated from the simulation model, for example:

    (a) Scientific simulations usually use a concept of *probes*, i. e., different types of views on the simulation data such as spatial visualization, trajectories or charts, and *controllers*, i. e., user interaction tools (sometimes called probes, too) [157]. Probes and controllers are logically neither part of the simulation engine nor of the simulated world.

    (b) More sophisticated functionalities similar to probes or controllers can be logically separable from a simulation model. For example, an "experiment generator" object might be useful to restart a setting several times with different preconditions.

    (c) Sometimes it can be desirable to separate functionalities from a simulation model although they, to some degree, logically belong to the simulated world. For example, in an evolutionary learning context it can be desirable to implement the learning functionality as a module which can be activated or unplugged to run the simulation in a non-learning mode.

    (d) Scheduling is a functionality that is usually separated from both the simulation model and the simulation engine.

2. Some simulation programs provide an API for runtime control of simulations from the "outside". For example, the traffic simulation AIMSUN offers such an API by informing the outside program of every step of the simulation providing methods for observation and interaction [8, 22]. This type of API requires the same type of control over a simulation run as the above-mentioned functionalities do.

3. Functionalities that are candidates for the SPL tend to be useful in many different settings. For example, a learning module can be used in several types of environments; as well, a scheduler is usually not bound to a specific type of environment; and most notably, probes usually can be interchanged between different simulation models.

In most common ABS frameworks, the issues listed above are resolved separately from each other. As an example, probes are usually implemented as a concept limited to observation and, possibly, user interaction; in MASON probes (called "inspectors" there) are specific GUI panels which allow the user to inspect or modify object parameters using reflections and Java Beans [117]. The simulation Swarm provides a more powerful probe concept using the reflection capabilities of Objective-C to allow any object's state to be read or set and any method to be called in a generic fashion [31, 129]. NetLogo provides a rather static probe capability by a set of GUI items that can observe and manipulate variables of predefined types [174]. However, in all these cases probes are solely meant to cover a subset of the functionalities listed in (1) and particularly not to implement schedulers or APIs (1d and 2).

In contrast, EAS treats all these functionalities in the same standardized way by using the SPL. For this purpose, AIMSUN's API concept has been generalized to provide a common interface for a large set of functionalities. In AIMSUN, a set of listeners is repeatedly notified about any occurring state changes in the simulation during a run. Each notification is passed to the outside program which is subsequently allowed to observe the current state and change it by performing specific method invocations. In EAS, this strategy has been applied to any interaction of the simulation engine with other parts of the simulation, particularly the simulation model.

The basic idea is to prohibit a direct communication of the simulation engine with the simulation model, but to introduce a layer of "listeners", called *plugins*, which are the only objects getting notifications about events or the proceeding of time from the simulation engine, cf. Fig. 3.8. The simulation model can be informed about notifications, only if they are passed to it by a plugin; such a notifying plugin is called *scheduler* in EAS. The plugins within the SPL are free to decide what to do when they receive a notification. If they pass it to the simulation model they serve as part of scheduling, otherwise they can perform any other activity such as visualizing the environment, managing user interactions, logging, API, and all the above-mentioned functionalities.

To facilitate this, every simulation model provides a set of methods that are accessible by each plugin and allow to pass notifications, observe the simulation state or control specific parts of the simulation. These methods are part of the implementation of an environment and can be flexibly programmed according to the requirements given by the intended purpose of the simulation. The simulation engine, on the other hand, provides a fixed set of methods which plugins can use to request notifications in order to be informed about specific time instants or events. The simulation engine ensures that every requested notification is accomplished at the according moment in simulation time. For this purpose, finally, each plugin also provides a fixed set of methods for receiving notifications, which can be called by the simulation engine. In a second step these notifications can be passed by the notified plugin to environments or agents, i. e., the simulation model (usually the main runnable). The simulation model, in turn, is responsible for the actual activities of objects or agents in the simulated world.

Therefore, plugins are the controlling interfaces between the course of time, which is maintained by the simulation engine, and a simulated environment, which is held in the simulation model. Particularly, there can be no activity at all in the simulation model if no plugins (more precisely, no schedulers) are involved. Consequently, at least one specific plugin called *master scheduler* is part of every simulation run and defines its basic scheduling. A master scheduler has, in addition to its plugin functionalities, the capability to create a simulation model in its initial state at the beginning of a run. Therefore, a master scheduler has influence on every aspect of a simulation from its creation to its termination. Particularly, it defines the basic relation of the course of time to the simulation model by requesting the points in time and types of events to be notified about, and passing notifications to the simulation model. (Although a master scheduler should provide for the basic scheduling of a simulation, there are, in principle, no objections to using additional schedulers; however, as they can be detached from the simulation – as opposed to the master scheduler, they should not be indispensable for the simulation's functioning, see below.)

All plugins, except the master scheduler, can be flexibly attached to or detached from a simulation run providing for a clean way of including and excluding functionalities that are optional to a simulation run. Moreover, the same plugin can be used in the scope of different simulation models to provide the same functionality to a variety of scenarios. For example, an evolutionary learning plugin can be used for a variety of different environments or agents to perform basic evolutionary operators during a run. Another example of a particularly universal plugin is a *visualizer* (called "video-plugin" in EAS) which has the purpose of depicting the state of an environment during simulation time. Fig. 3.3 as well as Figs. 3.5, 3.6 and 3.7 show screenshots of this plugin in EAS. It can be used to generate a live view or a pdf file from the simulation during a run, or to store a simulation run as an AVI film or a GIF animation.

3.2.4 The SPI Architecture from an MVC Perspective

The SPI architecture can be seen as a generalization of the well-known MVC architectural pattern. MVC, first described by Trygve Reenskaug in 1979 [162, 163] and implemented by Jim Althoff for Smalltalk-80, separates user interaction from data processing, letting both be altered independently [21, 112]. Beside the AIMSUN API concept, the MVC pattern has been the main inspiration for the SPI. Particularly, the visualization and user interaction parts of the EAS Framework are implemented according to the MVC pattern. MVC as well as its implementation in the EAS Framework by using the SPI is depicted schematically in Fig. 3.4. Within the MVC paradigm, techniques have been developed to handle, for instance, multiple views of an object, a single view that captures aspects from different objects, various controller types working simultaneously, or models physically separated from views and controllers, for example in web applications. These techniques can also be applied to the SPI making it possible

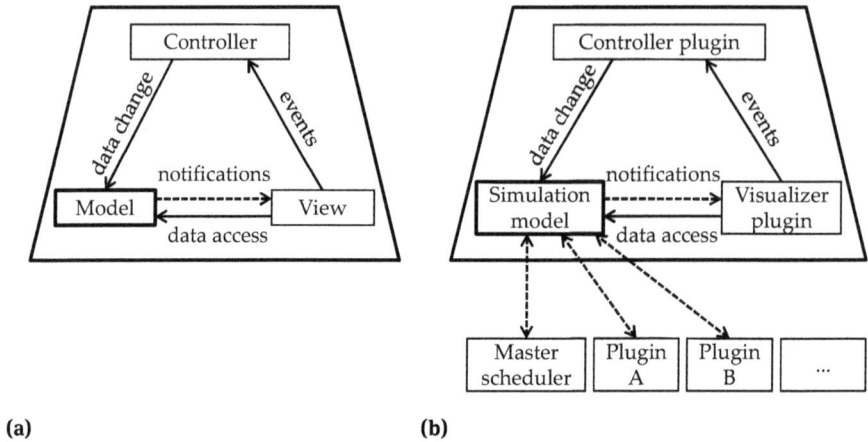

(a)   (b)

Fig. 3.4. Schematic views on the MVC architectural pattern. (a) Classic MVC pattern. (b) MVC pattern implemented within the SPI structure as a special use case of plugins; more plugins may be connected as indicated by the boxes at the bottom.

to flexibly attach and detach views and controllers at runtime and make them completely independent of the simulation model. However, the SPI goes beyond the idea of just separating specific functionalities, i. e., views and controllers, from the simulation model. Rather, if desired, virtually any functionality may be extracted from the simulation model and implemented as a plugin.

In MVC, the central unit of data processing is called *model*, which can be an arbitrary object in terms of object-oriented programming. The model is observed by one (or several) *view* object(s) which provide a visualization of the model, but cannot perform alterations to it. In principle, the model does not even have to be aware of the view, however, in new versions of the pattern, the model is also allowed to actively notify the view(s) of changes which can affect the visualization. User interactions are processed by one (or several) *controller* object(s). A controller can perform alterations to the model and communicate with its associated view using *events* (note that the term has a different meaning in the MVC pattern than defined above). There, a classic postulation has been that every view is associated with a unique controller and vice versa [21]. For the SPI, this limitation has turned out to be impractical: first, within the greater scope of functionalities covered by the SPI, such constraints seem rather artificial and would have been hard to specify; second, as the SPI is intended to be well-suited for unskilled programmers, any avoidable architectural dependencies have been omitted. Therefore, a single view plugin may be associated with multiple controller plugins as well as a single controller plugin may be used to interact with multiple view plugins. In EAS, the "videoplugin" serves as a controller and a view simultaneously, see below.

Ticks: 1512	AgentSmartCar: 0
Agent list	GetPriceSignal: 10.0
AgentSmartCar – ID: 0	GetPVValueGeneral: 0.0
AgentPriceSignal – ID: 1	IsLoading: true
	LoadIfCheap <Actuator>

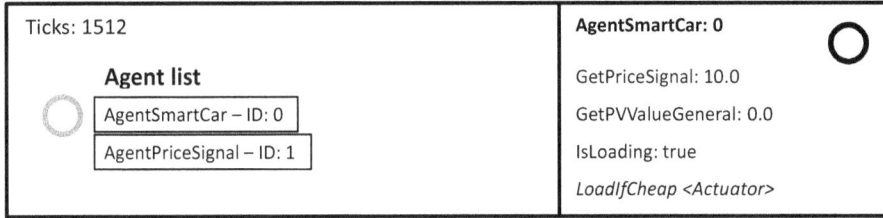

Fig. 3.5. Screenshot of a very basic abstract environment in EAS. The visualization is performed by the "videoplugin" which paints the environment view on the left side and the view of the selected agent (agent 0) on the right side. The videoplugin has additional capabilities such as zoom, rotation, translation etc. when a spatial environment is visualized. Furthermore, due to the intrinsic implementation of the `generateOutsideView` method, the plugin can generate movies (gif and avi) or pdf images (cf. Fig. 3.6) from any simulation.

Another violation of the classic MVC pattern exists within the EAS Framework which, however, is implementation-specific and not inherent to the SPI. Similar to the `toString` method in java, there is a method `generateOutsideView` in every agent or environment (more generally, in every `EASRunnable`) in EAS which provides a basic visualization functionality of that object. Allowing a visualization method within the simulation model violates a strict implementation of the MVC pattern. However, this practice is not supposed to support a mix-up of model implementations and view implementations. The method is rather meant to be used in a very basic way to provide for a visualization of any model even in the earliest developmental stages. In fact, the `generateOutsideView` methods predefined on the top-most levels of the class hierarchy in EAS provide for a sophisticated visualization of most of the supported simulation types, meaning that end user programmers in most cases do not have to implement anything concerning these methods.

For example, at the top environment level (class `AbstractEnvironment`) the `generateOutsideView` method is implemented to return an image of all agents in the environment drawn one above the other. Accordingly, at the top agent level (class `AbstractAgent`), the method generates an image of a string containing the id of the agent as well as its sensory perceptions and a list of its actuators. A screenshot of the resulting image visualizing a simple environment is shown in Fig. 3.5. At the level of 2D simulation (`AbstractEnvironment2D`), the method generates an image of the environment including all agents (`AbstractAgent2D`) that can be adapted by translation, rotation and zoom parameters. Fig. 3.6 shows a pdf extracted from a physics 2D simulation. Fig. 3.7 shows a 3D physics simulation based on the *Lightweight Java Game Library (LJGL)* of a cloth falling on top of a sphere. At this level, the 3D engine of the LJGL embedded in EAS is responsible for generating the environment view.

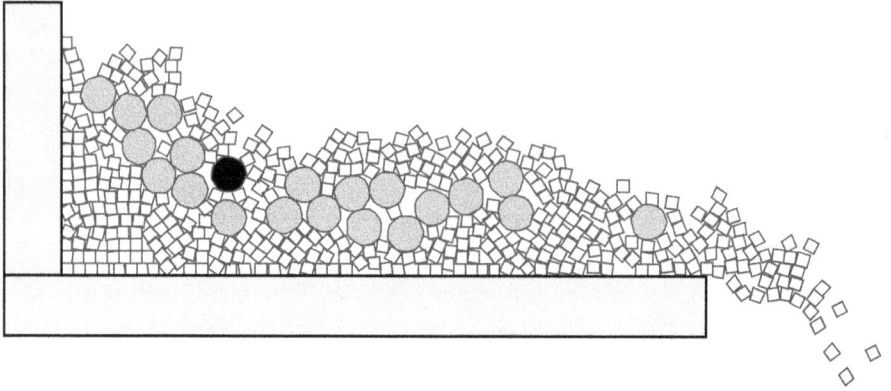

Fig. 3.6. Generated pdf of the visualization of a 2D physics environment. The environment contains several boxes and circles (modeled as agents) that have a mass and are affected by gravitation. The pdf is generated using a standard feature of the videoplugin.

3.2.5 Comparison of the SPI Architecture with State-of-the-Art ABS Frameworks

The proposed SPI architecture is in several respects similar to architectures used in state-of-the-art ABS frameworks such as MASON or SWARM. A general claim at design time has been to keep the simulation core as slim as possible to provide an error-free and computationally efficient foundation for the simulation program. Therefore, inspired by the MASON implementation, the EAS simulation engine consists basically of a loop which repeatedly produces notification messages based on a list of requests scheduled at specific time instances or events (cf. Sec. 3.3, class `SimulationTime`). As in MASON, there are objects that build entrance points to the simulation model by being notified about time instants and events, and progressing in time accordingly. In EAS, these objects are called *EAS runnables*, and one of them is called *main runnable* being referenced directly by the master scheduler as a receiver of notifications. A main runnable may contain further EAS runnables in a hierarchical way similar to the SWARM concept [31]. In principle, the hierarchy may even be broken by letting a runnable contain other runnables of its own type or of a supertype of that. In the rather straight forward ABS case, the main runnable represents an environment that contains arbitrarily many runnables representing agents. In MASON, the according type of objects is called "Steppable". The different nomenclature is intended to accentuate the ability of an EAS runnable to distinguish continuous time notifications from time-less events and ticks.

To the above-described extent, the architecture of EAS is similar to that of MASON or SWARM. The novel idea of the SPI architecture is to build an SPL as an intermediate tier between the simulation engine and the EAS runnables, cf. Fig. 3.8. One of the SPL's functions is to handle the propagation of notifications from the simulation engine to

Fig. 3.7. Visualization of a 3D physics environment implemented using the LJGL. The environment contains a sphere and a cloth snuggling against it while being pulled down by gravitation. Both cloth and sphere are implemented as EAS agents.

the simulation model. There, notifications sent to the SPL by the simulation engine are passed by plugins to the simulation model using an EAS runnable as entrance point. In this sense, the plugins act as a detour for information which, e. g., in MASON, would have been sent directly to the simulation model. By using this method, the SPL or, more precisely, a plugin in the SPL can provide scheduling for the simulation model by deciding which notifications are passed to it in which order. This mechanism is sufficiently powerful to make simulations which use the full SPI concept completely schedulable solely by plugins.

Current state-of-the-art ABS frameworks handle interactions of the simulation engine with the simulation model in a more direct way by calling methods from one side to the other. However, this type of architectures is not distinct from the SPI, but rather it constitutes a generalization of the SPI by allowing for certain direct references between objects that are not allowed in the SPI. Therefore, most current ABS frameworks can be subsequently upgraded to implement an SPI, by not forfeiting any of their original functionalities.

A basic SPL can be implemented very easily within most existing simulation platforms, without noticeably increasing runtime. In EAS the SPL holds a list of objects of the type `Plugin` which is a rather simple Java interface described in detail in Sec. 3.3. The list may contain arbitrarily many plugins simultaneously; in case of several plu-

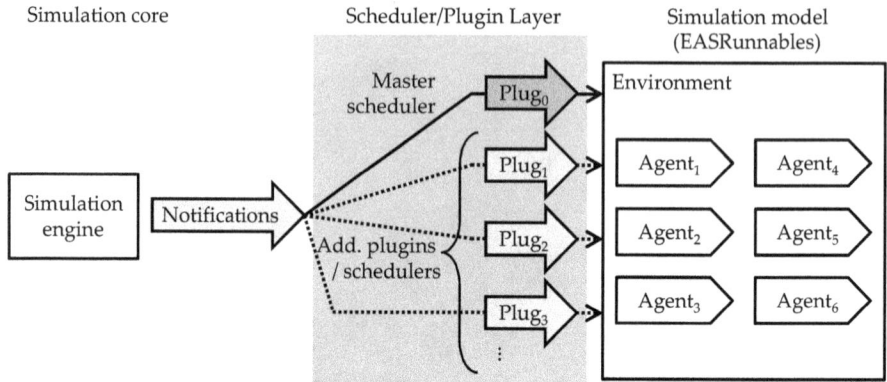

Fig. 3.8. Schematic view on the SPI architecture. Notifications are sent from the simulation engine to the SPL. The plugins in the SPL can pass notifications to the simulation model. One distinguished plugin called *master scheduler* provides for the basic scheduling of the model.

gins having to be notified at the same time, the notification order can be determined arbitrarily, for example according to the order in the list (as is done in EAS). At any notification, the simulation engine passes to the plugin to be notified a reference to the main runnable (by invoking the according method of the plugin). The plugin, in turn, can notify the main runnable or subordinate EAS runnables, if desired, by using the received reference. In principle, no more implementations are required to accomplish a basic version of the SPI. The next section describes the implementation of the SPI in EAS in more detail including some useful extensions to the basic version.

3.3 Implementation of the SPI within the EAS Framework

This section describes the implementation of the most important parts of the SPI architecture within the EAS Framework as well as some of its additional features. The description covers all classes belonging to the simulation core and several other classes, omitting implementation details where not required for understanding the basic concept. Furthermore, instructions are given on how to implement an SPL according to the SPI architecture within an existing ABS framework. The EAS version used as a reference is release 0_15_2013-10-11 which can be downloaded including all sources and documentation from sourceforge[1] [101].

---

1 http://sourceforge.net/projects/eas-framework

3.3.1 Overview

Overall, EAS (in the referenced version) contains 202 packages with 1,042 Java classes and approximately 105,000 *Lines Of Code (LOC)*. However, a major part of this code belongs to end-user implementations of simulation models for concrete scenarios. The fairly slim simulation core has about 500 LOC and consists of the classes

- `eas.simulation.SimulationTime` – implementation of the simulation engine,
- `eas.simulation.EASRunnable` (interface) – definition of the structure of entrance classes to the simulation model,
- `eas.simulation.Wink` – implementation of objects storing a time instant,
- `eas.plugins.Plugin` (interface) – definition of the structure of plugins, and
- `eas.plugins.MasterScheduler` (interface) – definition of the structure of master schedulers, a supertype of `Plugin`.

The class `SimulationTime` is an implementation of the simulation engine which is uniquely instantiated for every simulation run. Anytime a notification is requested, the `SimulationTime` object creates a `Wink` object which holds information about the current point in simulation time. This object, as well as optionally information about the event that triggered the notification, is sent to all listening objects of the types `Plugin` or `MasterScheduler`. From there, the plugins can pass the notification to all listening objects of the type `EASRunnable`. The following two methods of `EASRunnable` build the entrance point to the simulation model:

- `step(Wink)` for receiving notifications about time instants, and
- `handleEvent(EASEvent, Wink)` for receiving notification about events.

Fig. 3.9 depicts the interconnection of the core classes of EAS including the two classes `AbstractAgent` and `AbstractEnvironment` required for agent-based simulations. In the following, the implementation of these classes is explained in more detail.

3.3.2 Plugins

The interface `Plugin` defines the main functionality of all plugins in EAS. The most important methods of the interface are

(1) `void runBeforeSimulation(EASRunnable, ...)`,
(2) `void runDuringSimulation(EASRunnable, Wink, ...)`,
(3) `void runAfterSimulation(EASRunnable, ...)`, and
(4) `void handleEvent(EASRunnable, Event, Wink, ...)`.

As their names suggest, the first three are called by the `SimulationTime` object (1) before the beginning, (2) at runtime, and (3) after termination of the simulation, re-

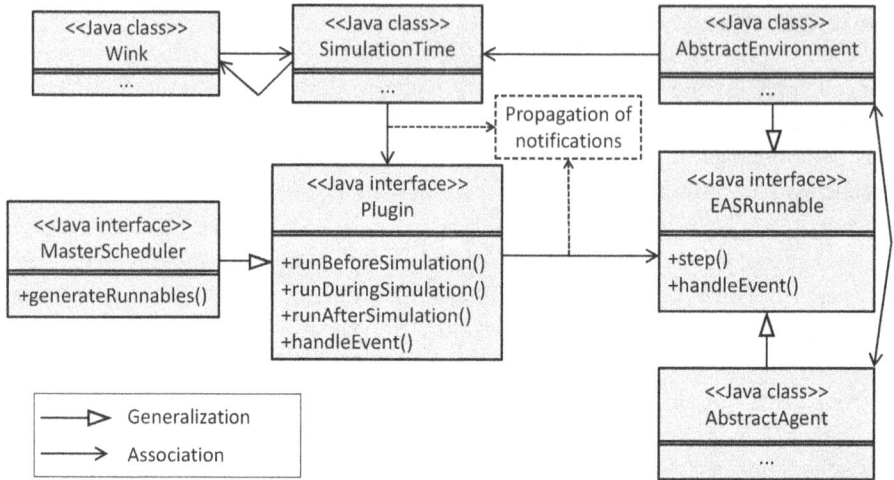

Fig. 3.9. Schematic view of the major classes of EAS including basic methods. Notifications about events and the proceeding of time are sent by class SimulationTime and reach the EAS runnables by making a detour over a plugin (usually a master scheduler).

spectively. All of them receive an EAS runnable object which is a reference to the main runnable of the current simulation run as defined by the master scheduler. Method (2) is called at each notification the plugin is registered for during a run. It additionally receives a Wink object which denotes the current time instance at notification time. Within this method, the step method of the received main runnable can be called to pass the notification to the simulation model. The method handleEvent (4) is invoked any time an event occurs that a plugin has registered for. It receives a reference to the main runnable, an event description and a Wink object denoting the last time step before the event has occurred (as events by themselves are time-less). All four methods receive one additional parameter concerning program parameters, as indicated by the three dots in the method signature, which is not relevant to the basic description here.

Plugins can register for and unsubscribe from notifications to determine what notifications to receive. For this purpose, the class SimulationTime contains several "request..." and "neglect..." methods as described below. By default, plugins are only registered for ticks, and they cannot prohibit being notified at the beginning and after termination of a simulation run. Furthermore, plugins can register for specific types of events using the interface EventFilter. By default, plugins are unsubscribed from all events.

3.3.3 Master Schedulers

Derived from `Plugin`, the interface `MasterScheduler` offers one additional method to allow for creation of a simulation's initial state:
(5) `EASRunnable[] generateRunnables(...)`.

Before the beginning of a run, this method is used to create one or more main runnable(s). If one main runnable is created (i. e., the returned list contains exactly one `EASRunnable` object), a single run is initiated by creating a `SimulationTime` object and all the according `Plugin` objects, and starting the simulation within the `SimulationTime` object. If several main runnables are created, each of these is attached to a specific `SimulationTime` object as well as to a copy of every plugin object including the master scheduler. At simulation start all the `SimulationTime` objects are triggered simultaneously, each running in parallel in different threads and accomplishing the notifications for its own list of plugins.

3.3.4 The classes `SimulationTime` and `Wink`

The class `SimulationTime` contains the main simulation loop, and it is responsible for sending notifications to plugins. Every plugin is notified at least twice, once before the start of the simulation by calling its `runBeforeSimulation` method, and once after termination by calling its `runAfterSimulation` method. Additionally, the plugin can be notified arbitrarily many times in between using the `runDuringSimulation` and `handleEvent` methods. Both, these methods receive a `Wink` object which is generated by the `SimulationTime` object to contain the current time instant at notification time or the last time instant occurred before an asynchronous event has been triggered.

To customize the plugins' notification policies, the class `SimulationTime` provides the methods listed in Tab. 3.1. Using these methods, a plugin *P* can choose which types of notifications to receive in the future. Each of the methods in the list can be called without a plugin parameter P, too, affecting in that case notifications of the master scheduler. Note, that the method `requestAllNotifications` informs a plugin about all time steps that have been requested by any other plugin as well as all ticks and all events. The complementary method `neglectAllNotifications`, on the other hand, does not lead to that plugin never being notified again (this functionality is rarely needed and implemented in a method called `neverNotifyAgain`). Rather, it stops sending all notifications to the plugin and recreates the notification policy which was valid before the method `requestAllNotifications` has been called (or it has no effect if the method has not been called before).

Table 3.1. Methods for customizing plugin notifications in class SimulationTime. The table shows the effects on the notifications of a Plugin P (which can be a master scheduler) after invocation of the methods denoted in the first column. There, $t \in \mathbb{R}$ is an arbitrary time instant, and F is an event filter. (Cf. description in text for explanation of method `neglectAllNotifications`.)

Invocation of method	Time step t	Tick ($t \in \mathbb{N}$)	Event matching F
[standard, no invocation]	no	yes	no
`requestTicks(P)`	[no change]	yes	[no change]
`neglectTicks(P)`	[no change]	no	[no change]
`requestNotification(P, t)`	yes	[no change]	[no change]
`neglectNotification(P, t)`	no	[no change]	[no change]
`requestAllNotifications(P)`	[if t req. by any plugin]	yes	yes
`neglectAllNotifications(P)`	[if req. before]	[if req. before]	[if req. before]
`requestEvents(P, F)`	[no change]	[no change]	yes
`requestAllEvents(P)`	[no change]	[no change]	yes

3.3.5 The Interface `EASRunnable`

The interface `EASRunnable` has to be implemented by every object that is supposed to be run by the simulation engine; in the agent-based case, this concerns environments and agents. For the purpose of controlling the simulated object's progression in time and its reaction to events, the interface contains the methods `step(Wink)` and `handleEvent(EASEvent, Wink)`. The `step` method defines how the object changes due to the course of time by performing a state change for any received notification depending on both its current state and the current simulation time. The method `handleEvent` defines how the EAS runnable reacts on events.

The EAS runnables' `step` and `handleEvent` methods can be invoked by the master scheduler of a run, but also by other plugins or EAS runnables. For example, an environment usually acts as main runnable which contains a list of agents as subsequent EAS runnables whose `step` and `handleEvent` methods can be invoked by the environment itself.

3.3.6 "Everything is an Agent": a Philosophical Decision

In EAS, every instance of the EAS runnable type `AbstractAgent` or any class derived from it, is called *agent*. Particularly, the class `AbstractEnvironment`, which is the origin of all environments, is itself derived from `AbstractAgent`. Therefore, technically all agent-based simulations in EAS are solely based on agents and their interactions between each other. Especially, in a very recursive manner, agents can contain other agents, and consequently environments can contain other environments or agents as parts of themselves. This conceptual decision is on the one hand convenient from an implementation point of view as environments and agents can be treated equally and interact in a flexible way. On the other hand, and more importantly, the agent metaphor that arises from the idea that "everything is an agent" makes sense in connection to real-world applications. If the concepts of *environments* and *agents* are treated differently, it can become a truly philosophical question which aspects of the world to view as parts of the environment and which as parts of an agent (cf. the work by Weyns et al. on the topic of environments in multi-agent systems [202, 203, 204]).

When simulating robots, for example, a first approach might be to represent a whole robot, i. e., the physical unit with all its arms, legs, wheels and sensors, as an agent. However, in a simulation with a sufficiently high degree of detail immediately serious questions can arise, for example, what happens if the robot loses a leg? Is the robot from that point on divided into two agents? Or is it one agent that shrinks or grows in size as its body moves toward or away from the leg? Or does the leg now belong to the environment and no longer to the agent – has it switched its state from being active to being passive? And if so, what exactly would that imply for its future treatment? To avoid these inconsistencies, another idea might be to treat all physical parts of an agent as part of the environment, while solely its controller as some kind of inner decision-maker or "soul" is defined as the actual agent. But looking more closely even this extreme point of view does not offer a consistent model. In a detailed simulation, the controller might be implemented in a way to have a physical body or "brain" that is subject to the laws of physics of that environment just as the arms and legs of a robot are, therefore only shifting the above issue of a lost leg to the question of what happens when the controller is physically changed by the environment. And even under the assumption that the controller is unaffected by the laws of the environment, the question remains how to connect it with the environment, i. e., what to do with sensory (or actuator) stimuli. Usually, sensory stimuli come from the environment, so it might seem reasonable to define them as part of the environment. But what if the internal state of a robot or even the controller itself are perceptible by a robot's controller. In the case of MAPTs, which are part of the controller structure in Chap. 5, the controller can "perceive" aspects of its own past execution (just as a brain's functioning depends on its own past states). Doesn't this imply viewing the controller including its method of perception as part of the environment, too? (And what, then, is the inner "soul" that can legitimately be called "agent"?)

Any of the suggested solutions seem unsatisfactory, and the question itself is, in its essence, strongly connected to the philosophical or biological issue of defining what a living being is made of and how it is separated from its surrounding environment (cf. the complex, but intriguing answer given by Maturana and Varela [125]). From a materialist's perspective such a distinction may not even make any sense. From any other perspective it is a difficult problem to define the connection of the "immaterial" world of perception and decision making with the material world surrounding it. In most agent simulations the end user is forced to resolve this issue when implementing a simulation scenario.

Treating the environment itself as an agent (or, to the same effect, the agent as a type of environment) reduces the difficulty of deciding how agents should be separated from the environment to a (from a software developer's point of view) more pleasant question of how the different objects to simulate should be distributed among different types of agents. In fact, the question becomes much like the well-known problem of any object-oriented software development process, to decide how to structure the classes among each other such that the resulting objects and their interactions make most sense. In the above example, a robot's leg, as any object to simulate, would be implemented as an agent that is part of a larger agent, the robot, which is part of the environment, an even larger agent. If the leg is lost by the robot, it remains the same type of agent as before, and only changes its mode of interaction with the other agents.

Given this example, it can be argued that viewing the environment as an agent does not affect the core of the above-mentioned problem as there is still an agent called "environment" and it is not clear which parts of the world to simulate directly belong to it and which parts should be parts of other sub-agents within that environment agent. However, the great advantage of a unified agent/environment view is that decisions made in the beginning of the software development process are not fixed, but can usually be changed easily at any time if a better way of structuring comes up. In other words, treating agents and environments uniformly reduces the architectural decisions required by the end user from a philosophical to a pragmatic level.

3.3.7 Running a Simulation

To start a simulation, the simulation engine, i. e., the class SimulationTime in EAS, has to enter the main simulation loop. For the purpose of standardizing this process, a class SimulationStarter has been implemented which is used to perform all necessary preparations and start a simulation. It provides a main method which initiates the following five-step process:

1. The master scheduler is selected and constructed.
2. The main runnables $R_1, \ldots, R_n$ ($n \in \mathbb{N}$) are generated by using the master scheduler's generateRunnables method.
3. A set of SimulationTime objects $T_1, \ldots, T_n$ is created, and every T_i is associated to R_i ($i \in \{1, \ldots, n\}$).
4. For the selected plugins $P_0, P_1, \ldots, P_m$ ($m \in \mathbb{N}_0$; P_0 being the master scheduler), a list of copies $P_{0_i}, \ldots, P_{m_i}$ ($i \in \{1, \ldots, n\}$) is created, and the list $(P_{0_i}, \ldots, P_{m_i})$ is associated to R_i. Therefore, every main runnable is associated to an individual set of copies of the SimulationTime and Plugin objects. This procedure allows for each main runnable to be treated separately including all according agents and plugins.

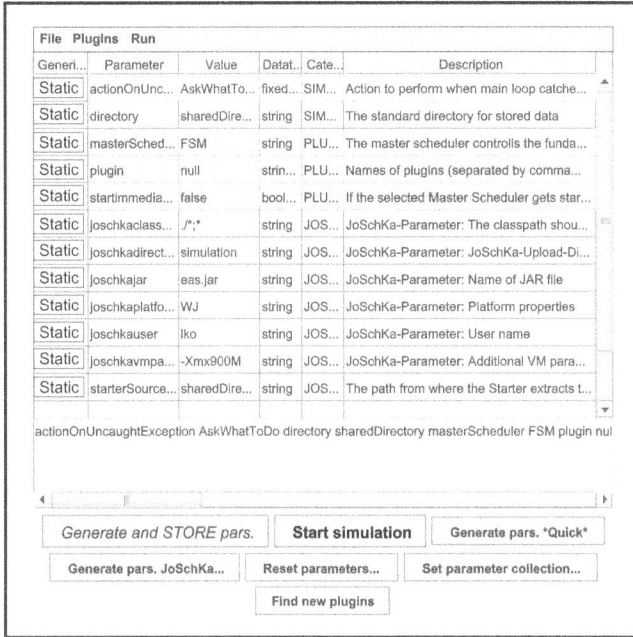

File	Plugins	Run				
Generi...	Parameter	Value	Datat...	Cate...	Description	
Static	actionOnUnc...	AskWhatTo...	fixed...	SIM...	Action to perform when main loop catche...	
Static	directory	sharedDire...	string	SIM...	The standard directory for stored data	
Static	masterSched...	FSM	string	PLU...	The master scheduler controlls the funda...	
Static	plugin	null	strin...	PLU...	Names of plugins (separated by comma...	
Static	startimmedia...	false	bool...	PLU...	If the selected Master Scheduler gets star...	
Static	joschkaclass...	./*;*	string	JOS...	JoSchKa-Parameter: The classpath shou...	
Static	joschkadirect...	simulation	string	JOS...	JoSchKa-Parameter: JoSchKa-Upload-Di...	
Static	joschkajar	eas.jar	string	JOS...	JoSchKa-Parameter: Name of JAR file	
Static	joschkaplatfo...	WJ	string	JOS...	JoSchKa-Parameter: Platform properties	
Static	joschkauser	lko	string	JOS...	JoSchKa-Parameter: User name	
Static	joschkavmpa...	-Xmx900M	string	JOS...	JoSchKa-Parameter: Additional VM para...	
Static	starterSource...	sharedDire...	string	JOS...	The path from where the Starter extracts t...	

actionOnUncaughtException AskWhatToDo directory sharedDirectory masterScheduler FSM plugin nul

Generate and STORE pars.	Start simulation	Generate pars. "Quick"
Generate pars. JoSchKa...	Reset parameters...	Set parameter collection...
	Find new plugins	

Fig. 3.10. Screenshot of the starter GUI. The program parameters are shown in the upper part of the window. Plugin-related parameters are highlighted (i. e., the parameters for choosing a master scheduler and a list of plugins for a run, "masterScheduler" and "plugin", as well as a parameter called "startimmediately" which, if set to *true*, omits showing the starter GUI, immediately running the simulation on program start). The parameter values can be changed and stored in a parameter file or as a JoSchKa package.

5. Each of the `SimulationTime` objects T_i is caused to enter the main simulation loop in a separate thread. During the run its own copies of the selected plugins $P_{0_i}, \ldots, P_{m_i}$ are notified.

The `main` method of the class `SimulationStarter` has to be started with program parameters specifying the master scheduler and other plugins as well as additional parameters. The command line parameters given to this method as arguments as well as a parameter file called `parameters.txt` in the main directory are used to create this collection of program parameters. Beyond that, EAS provides a mechanism to create new parameters flexibly corresponding to each plugin, to make the plugin behavior adjustable at program start. For example, a plugin parameter may specify the number of agents to be constructed by a specific master scheduler. To simplify the process of selecting parameters, there is a class `Starter` provided which includes a *Graphical User Interface (GUI)* guiding the selection of plugins and other program parameters as well as allowing to start a simulation, cf. Fig. 3.10. Furthermore, the starter GUI allows to specify a collection of job packages that can be stored in the JoSchKa file format to be distributed on computer pools or in a cloud.

These and other additional functionalities of EAS, for example, the sensor/actuator concept and the implementation of agent brains, go beyond the scope of this thesis. However, they are described in an extensive Javadoc package as well as in a short tutorial in German which both are part of the EAS Framework.

3.3.8 Getting Started

As stated above, EAS can be downloaded from sourceforge including all sources [101]. Other than the EAS packages, Java 7 [183] or higher is required to run simulations, and the use of an Eclipse IDE [41] is recommended to implement simulations as EAS comes with pre-built Eclipse project settings. To start developing an own agent simulation in EAS, typically three steps have to be done:

1. Create an environment class derived from `AbstractEnvironment` (for spatial representation: `AbstractEnvironment2D`, `AbstractEnvironment3D`, ...).
2. Create one or more agent class(es) derived from `AbstractAgent` (for spatial representation: `AbstractAgent2D`, `AbstractAgent3D`, ...).
3. Create a master scheduler derived from `MasterScheduler` or, more typically, `AbstractDefaultMaster` (a helper class which implements scheduling behavior as required in the most cases); let the `generateRunnables` method create an instance of the new environment and insert agents as desired.

The abbreviation "EAS" can be used as a mnemonic aid to remember the recommended implementation sequence "Environment → Agent → Scheduler".

For a first start, when using `AbstractDefaultMaster`, the only real implementations, besides empty method or constructor bodies, have to be performed in the methods `generateRunnables` and `id` of the master scheduler. The former has to return an instance of an environment (or a list of several environments) including agents which can be inserted by the `addAgent` methods of the class `AbstractEnvironment`. The latter simply has to return an arbitrary identification string used to select the master scheduler at program start. EAS finds the newly generated master scheduler class automatically using reflections and makes its id appear in the list of master schedulers in the starter GUI after clicking on the "Find new plugins..." button. When working without the starter GUI, the list of plugins can be refreshed by deleting the file `plugins.dat` from the main directory before starting a simulation. After selecting the master scheduler id from the list of master schedulers and "videoplugin" from the list of plugins, the new simulation can be started and visualized by clicking on "Start simulation...".

Table 3.2. Summary of the 16 tasks of the Stupid Model. A detailed definition of the Stupid Model tasks is given in [156] (the summary is taken from [157]).

No.	Task
1	100 agents ("bugs") distributed randomly into a 100^2 grid (movement and visualization).
2	A second bug action: growing by a constant amount.
3	Habitat cells that grow food; bug growth is equal to the food they consume from their cell.
4	"Probes" letting the user see the instance variables of selected cells and bugs.
5	Parameter displays letting the user change the value of key parameters at run time.
6	A histogram of bug sizes.
7	A stopping rule that causes execution to end when any bug reaches a size of 1000.
8	File output of the minimum, mean, and maximum bug sizes each time step.
9	Randomization of the order in which bugs move.
10	Size-ordering of execution order: bugs move in descending size order.
11	Optimal movement: bugs move to the cell within a radius of 4 that provides highest growth.
12	Bugs have a constant mortality probability, reproduce when they reach a size of 10.
13	A graph of the number of bugs.
14	Initial bug sizes drawn from a random normal distribution.
15	Cell food production rates read from an input file; graphical display of cell food availability.
16	A second "species": predator agents that hunt bugs.

3.4 A Comparative Study and Evaluation of the EAS Framework

In this section, a study comparing EAS to the ABS frameworks MASON and NetLogo is presented. MASON has been chosen for its architectural comparability to EAS. MASON and EAS are both Java-based ABS frameworks and flexible enough to build various complex scenarios. NetLogo, on the other hand, has been chosen as it is often used by beginners due to its simplicity. The frameworks are compared to each other with respect to the objective of guiding unskilled users to well-structured code.

3.4.1 Method of Experimentation

The study has been conducted within the advanced laboratory course OCLR, cf. Sec. 3.1, with six students as test subjects during one semester (18 weeks), starting in October 2011. As a test case, the well-known *Stupid Model* [156] has been used which consists of 16 tasks to implement in a simulation program. In literature, the Stupid Model has been used several times for the evaluation of ABS frameworks, originally by Railsback et al. [157]. Tab. 3.2 shows a summary of the 16 tasks of the Stupid Model. Every test subject had to individually implement the complete Stupid Model using each of the three ABS frameworks where the implementation order was left to the test subjects. The test subjects were allowed to discuss implementation questions with each other and, depending on the ABS framework in question, with one of three instructors, each assigned to one of the ABS frameworks. To keep the results as unbiased as possible of the instructors' background knowledge, each instructor was assigned to an ABS program, he had no former experiences with. The instructors had a chance to become acquainted to their simulations by allowing two months of initial training before the beginning of the course. All instructors as well as all test subjects had earlier experience with Java. None of the test subjects knew MASON or EAS, and two of them had worked with NetLogo before. Other than that none of the test subjects had any experiences with ABS. After the implementation phase, the students were anonymously asked a series of survey questions. The complete survey including questions and results (in German) can be downloaded in SPSS and Excel format[2].

Additionally, a structural analysis of the LOC and the number of files (corresponding to classes in the Java case, as no inner classes have been used by any of the test subjects) in relation to the tasks of the Stupid Model has been performed. The number of files and LOC has been related to the structuredness of code. In *Object-Oriented (OO)* languages it is a desirable goal to keep the number of LOC per file low. In [11] a number as low as 50 LOC per class is stated as a maximum. The measure does not apply perfectly well to NetLogo as it is not OO, but file size indicates structuredness there, too. In NetLogo, a program is usually read and executed from a single file; however, some of the test subjects chose to introduce a second file at some point to improve structuredness by distributing the 16 subtasks on two places (nobody chose to use more than two files). There, one file was used for a subset of the 16 tasks while the other implemented the remaining ones. As a basic foundation of the model had to be present in both files, this procedure led to redundancy. In these cases, the number of files has been counted as two, but the duplicate parts of the code appearing in both files have been considered only once for the LOC count. A normal distribution has been assumed to underlie the data. *Standard Deviation (SD)* and Student's t-test have been calculated to indicate statistical significance.

The program versions used in the study have been 0.9 for EAS, 15 for MASON and 4.1.3 for NetLogo (note that a later version of EAS has already been available, but has not been used as some improvements have been performed after studying the Stupid Model specification, which possibly would have biased the results). Eclipse has been used as an IDE for the Java-based frameworks, NetLogo comes with its own IDE. The program `cloc`[3] has been used to count LOC.

---

2 http://www.aifb.kit.edu/images/0/0e/AllSurveyDataWSC2012.zip
3 http://cloc.sourceforge.net/

3.4.2 Results and Discussion

Overall, the test subjects have been able to complete 268 out of the $16 \cdot 3 \cdot 6 = 288$ required implementations (6 have not been completed for EAS and NetLogo, respectively, 8 for MASON). The incomplete cases are counted in the structural analysis by using the mean value of all the other test subjects for the according task and simulation.

Structural analysis of the code. A two-tailed t-test has been performed for all pairs of simulations on the mean summed-up LOC and mean summed-up number of files per test subject, respectively. It indicates that all the mean values are pairwise different with a probability of at least 99.9 % except for the LOC of MASON vs. EAS where the probability still calculates to more than 99.0 %. Therefore, in the following the structural differences observed between the simulations are considered significant. Fig. 3.11a shows the number of LOC/file for the three frameworks summed up over the 16 subtasks. All three simulations break through the bound of 50 LOC/file. However, overall the EAS implementations require the fewest LOC/file in comparison to the other two. The LOC/file seems to stagnate at below 70 for EAS, meaning that a further increasing number of tasks might not significantly increase the LOC/file count. This assumption has been examined by calculating the average over all EAS-based simulations implemented so far in the eas.users package. It contains code from various users, students as well as highly skilled developers, but excluding the code of this study. Overall this calculated to $\frac{27,670}{405} \approx 68.3$ LOC/file, cf. horizontal solid line in Fig. 3.11a. A similar test with MASON's sim.app package lead to $\frac{12,738}{165} = 77.2$ LOC/file. This data has to be considered with caution as the code has been produced at least partially by skilled users which is particularly not the case for this study. NetLogo's LOC/file count increases essentially linearly with the LOC, except for the infrequent distribution of code to two files.

Beside the LOC/file count, the absolute LOC count required for the three simulations has been studied, cf. Fig. 3.11b. NetLogo has required the fewest LOC (195; $SD = 12$), while EAS has required the most (856; $SD = 36.5$), and MASON has been in between (632; $SD = 49$). It is notable that for tasks 5, 6 and 13 the difference between EAS and the other two frameworks is considerably higher than for the other tasks. This can be related to the lack of chart and parameter changing support in the provided EAS version, which is crucial for these tasks. The according functionalities have been implemented in a later version of EAS. Ignoring these outliers, there is on average practically no difference between the Java-based frameworks in the total LOC count.

A related observation has been made in combination with the post-course survey. There, tasks including file operations (tasks 8 and 15) have been reported to be time-consuming when implemented the first time (task 8) in all frameworks (cf. Fig. 3.12); but the second time (task 15) time consumption has been reported to be significantly lower for EAS than for the other two frameworks, while the LOC count has been sig-

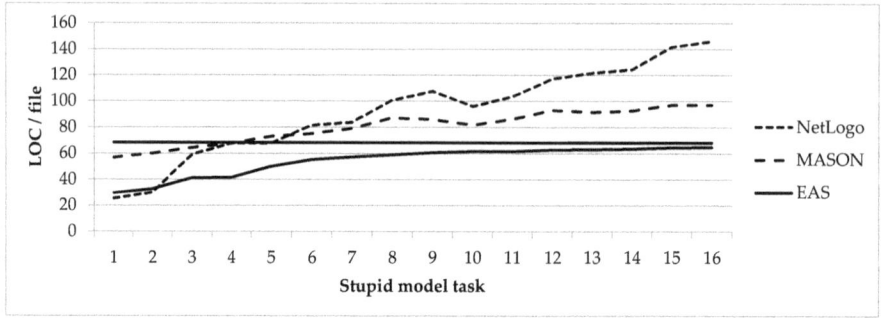

(a)

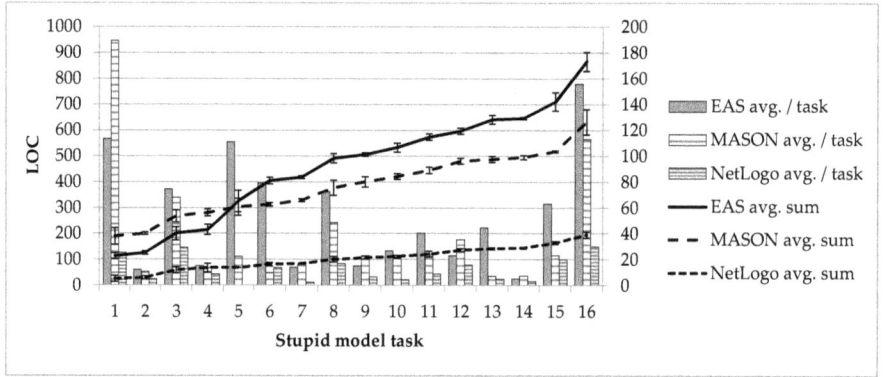

(b)

Fig. 3.11. Measures of LOC and file count for the three simulations. (a) Average LOC/file, summed up incrementally over the 16 Stupid Model tasks; horizontal line: average LOC/file in all classes of EAS users package. The LOC/file count for EAS seems to approach the value observed in the users package which is significantly lower than the values for both NetLogo and MASON. (b) Total lines of code for each of the 16 Stupid Model tasks individually (bars, right axis), and summed up incrementally (lines, left axis), respectively; error bars denote SD for LOC of individual tasks (not sum).

nificantly higher. This indicates that code has been copied and pasted and hints at a requirement for better file access functionality in EAS; this has been accomplished in a newer version. Overall, the implementation time for EAS has been reported to be higher than for the other two simulations, cf. discussion below. The test subjects attributed that to the lack of documentation and some types of high-level functionality. While such findings point at (minor) drawbacks of EAS as a whole, they have to be excluded when considering the ability of SPI-based simulations to create structured code.

In conclusion, it can be stated that EAS as a young ABS framework has not been as elaborate as MASON and NetLogo in terms of predefined functionality in the studied version, and that it has lacked documentation. Therefore, more LOC and in sum more time has been needed to implement the Stupid Model in EAS. However, the results indicate that the EAS architecture leads to code that is better structured and can be reused more easily than with the other two simulations. With respect to the SPI concept, particularly tasks 5, 6, 8, and 13 have been usually implemented as plugins, while tasks 9 and 10 have mostly been implemented using the scheduling capability of the master scheduler. For the first task, EAS offers a reusable plugin ("videoplugin") that can be attached to any simulation without requiring any implementations. NetLogo, too, does not require any special code for visualization. In contrast, MASON requires a considerable amount of code for this task which is mainly due to the rather complicated (although elaborate) implementation of visualization.

Evaluation of the surveys The main objective of the surveys has been to compare the times unskilled users need to implement a model in the different ABS frameworks, and to further examine the resulting implementations for their structuredness and reusability. The survey questions can be clustered in four main parts:
1. EAS-specific,
2. MASON-specific,
3. NetLogo-specific, and
4. Comparative.

Each of the first three parts has consisted of questions specifically related to one of the simulations as indicated by their names. They had to be answered right after completing an implementation of the Stupid Model in the according simulation. The questions have been identical in each of these three parts, except for the adaptation to the respective simulation. The fourth part has consisted of comparative questions that had to be answered after the completion of the implementation in all three simulations.

Unsurprisingly, the well-established programs MASON and NetLogo took the lead in the comparing questions of part 4, particularly concerning available documentation and examples. Furthermore, certain functionalities (mainly chart generation and parameter change at runtime, as well as file operations) have been reported to be straightforward in MASON and NetLogo, but rather complicated in EAS. The instal-

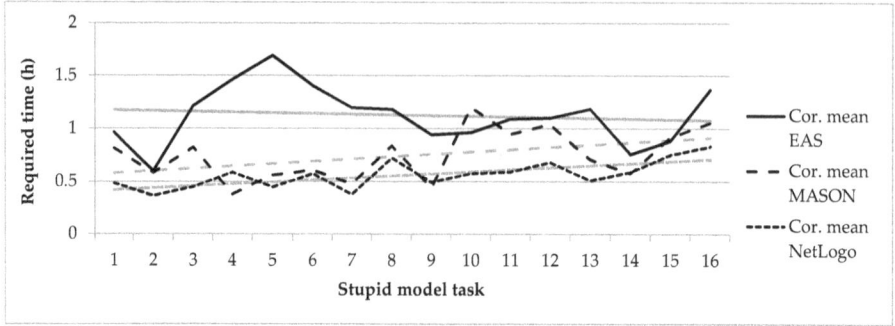

Fig. 3.12. Reported time required to implement tasks 1-16 of the Stupid Model. The time reported for EAS is overall considerably higher than for each of the other two simulations. However, the linear regression measures denoted by the gray lines show that the required amount of time continues to increase with the number of completed tasks for MASON and NetLogo, while it slightly decreases for EAS. This might indicate that the test subjects adjusted to the framework over time and learned to solve the tasks faster than in the beginning. It might also indicate that the structuredness of EAS leads to a better reusability of already implemented code. (Of course, due to the rather sparse data available, this may be wishful thinking.)

lation of the frameworks and the launch of provided examples have been reported to be easiest in NetLogo. Nevertheless, when being asked to choose a simulation framework for a later assignment of the OCLR course, EAS came of fairly well by being chosen twice as were MASON and NetLogo.

In one major block of the three simulation-specific surveys, the students had to specify the relative amount of time per framework they spent for completing each of the 16 tasks as well as the total time for the whole Stupid Model, cf. Fig. 3.12. To reduce subjective impressions, the students had to distribute 800 points in total to the 16 tasks with a maximum of 100 points per task. These points have been recalculated to the actual time consumption per task, by considering the reported total time. As the figure clearly indicates, visualization of charts has been the most time-consuming implementation in EAS. This is supported by the code analysis, too (cf. Fig. 3.11b), and led to the implementation of a major plugin supporting chart functionality ("chart-plugin") which is available since version 0.12 of EAS. Complex scheduling (task 10) in MASON seemingly has been more complicated to the test subjects than in any of the other frameworks, as they required a significantly greater amount of time for the MASON implementation (although the number of LOC has been greatest in EAS). This indicates that the scheduling capability embedded in the plugin structure of EAS is more intuitive to unskilled users than the implementation in MASON. The NetLogo scheduling implementations have required the least time and fewest LOC of the three, however, it has to be considered that NetLogo is not completely schedulable.

Overall, time and effort have been smallest in NetLogo, followed by MASON and then EAS. Still there is evidence that the costs for implementing features in NetLogo have been lowest for the first, simpler tasks and grew gradually with the rising complexity of the tasks. A similar development can be observed for MASON, while there is no such tendency obvious for EAS. This could indicate that the architectural structure of EAS supports additions and extensions to simulations more easily. However, it has to be considered that the Stupid Model is a very simple model even at the most complex subtasks, and that the results might not be generalizable to more sophisticated simulations. On the other hand, EAS has been extended after this comparative study to provide for the formerly lacking types of high-level functionality. Moreover, the available documentation and programming examples have been improved and are planned to be further developed in the future. Overall, EAS, and more generally the SPI have proven promising for providing a powerful and yet intuitive foundation for conducting complex simulation projects which include unskilled as well as experienced programmers. This statement is an honest opinion of the author, built – beyond the rather preliminary experimental data just presented – upon a profound experience with the framework in real-world circumstances.

3.5 Chapter Résumé

In this chapter an architectural concept for ABS frameworks called SPI has been proposed, and an ABS framework implementing the SPI called EAS Framework has been presented. The novel architecture is intended to improve modularity and structuredness as well as reusability of developed simulation code. There, a goal is to not only allow implementations to have these properties, but to actively guide particularly unskilled users in an intuitive way to a well-structured implementation. To achieve these objectives, the SPI introduces an additional layer between the simulation engine and the simulation model allowing to place all the functionalities there that are logically separable from the simulation model. This includes (but is not limited to) scheduling, probes, visualization, statistics calculations, logging, API to other programming languages etc. Therefore, the SPI offers a uniform way of extracting functionality from the simulation model which does not belong there, and implement it in a modular way at a standardized place. The extracted modules can be attached and detached flexibly, and used in different simulations making the code reusable.

The SPI has been evaluated by comparing EAS to the state-of-the-art simulation frameworks MASON and NetLogo. Six test subjects, most unfamiliar with any of the simulations, individually implemented the "Stupid Model" using all three simulations. The results indicate that the implementations in EAS are structured significantly better in terms of LOC/file (class) and reusability of the code than in both MASON and NetLogo. Moreover, the SPI concept has been credited by the test subjects as beneficial compared to the other simulations. However, the total LOC count is highest in EAS which can be partly explained by the higher number of classes required in EAS, but it certainly also has to be ascribed to the elaborateness of the two other simulations. Particularly NetLogo comes with a lot of pre-assembled components that can be integrated using a GUI without requiring any implementations (although being rather inflexible in return).

EAS has been improved after the comparative study from a functional point of view meaning that more standard functionality is provided now. This includes a chart plugin (particularly easy to use when combined with the "ObserverAndController" plugin), a plugin that allows changing program parameters at runtime, a runtime plugin manager, a trajectory plugin, improved visualization and better file access. Furthermore, the SPI architecture is not bound to the EAS Framework, but can be implemented fairly easily within most state-of-the-art simulation frameworks. Particularly a MASON version of the SPI has been theoretically developed and is planned to be implemented in future.

Overall, the results obtained from the comparative study and practical experiences with students and post-graduates using the EAS Framework for real scientific applications indicate that EAS offers several practical advantages in an academic environment. Moreover, the results show that the SPI can improve the structuredness and modularity of code in ABS simulations. As the architecture can be implemented within most state-of-the-art ABS frameworks, the EAS Framework is not necessarily required for programmers wishing to exploit the SPI concept.

The EAS Framework has been the foundation of all simulation experiments throughout this thesis. In the next chapter, it is used to evaluate an FSM-based evolutionary approach for learning behaviors in swarms of mobile robots.

4 Evolution Using Finite State Machines

Preamble:

> "*Beautiful is better than ugly.*
> *Explicit is better than implicit.*
> *Simple is better than complex.*
> *Complex is better than complicated.*
> *Flat is better than nested.*
> *Sparse is better than dense.*
> *Readability counts.*
> *[…]*
> *If the implementation is hard to explain, it's a bad idea.*
> *If the implementation is easy to explain, it may be a good idea."*

These postulations for computer programs are part of the Zen of Python *[150]. While it is widely recognized that human-designed programs should preferably follow these rules to keep programs comprehensible, automatically designed robot programs – whose correct functioning may be extremely crucial – still tend to be graded solely by observation. By accomplishing most of the above criteria, this chapter is intended to provide an evolvable controller model which can be formally analyzed to prove the fulfillment of various desired behavioral properties.*

The field of ER is a broad merger of various techniques for the development of robotic controllers inspired by principles found in natural evolution. Evolution has been proven capable of finding control systems which outperform manually designed solutions in terms of effectiveness in solving a desired task, simplicity of controllers and generalizability [14, 142, 199]. Although it is still an open question if truly complex behavior can be learned by the means of ER, the results achieved in the last two decades are promising and point toward possible breakthroughs in the future.

In most of the recently reported successful approaches, robots have been controlled by ANNs or similarly complex control structures as these are known to provide effective learning capabilities. However, due to the complexity of the evolved controllers, their quality cannot be evaluated automatically by a structural analysis. Rather, it has to be estimated by observing the resulting behavior in terms of fitness or "by instinct" which is infeasible for a large set of critical applications.

This chapter proposes an evolutionary approach based on MARB ("Moore Automaton for Robot Behavior") controllers, which implement a fairly simple and analyzable FSM structure. Overcoming the inflexibility of FSMs, in terms of adaptability in an evolutionary scenario, is a major topic of this and the following chapter. In this chapter, MARBs are used as a direct encoding for robot controllers meaning that there is no distinction between genotypic and phenotypic representations. A broad range of experiments in simulation and on real robots is presented showing that controllers of similar complexities as reported in current literature can be evolved. Several properties of MARBs including techniques to increase adaptability as well as an efficient encoding on real robots are discussed. Exemplarily, it is outlined how evolved controllers can be automatically analyzed using a simple state reachability criterion.

The theoretical foundations of this chapter have been published in slightly evolving versions in several papers [66, 102, 105, 106, 107]. The first description of a very early version of the model has been given in a 2007 diploma thesis [100], and the foundations of the real-robot model used in Sec. 4.4 have first been published in a 2011 diploma thesis [65]. The experimental results have been published in the above-mentioned papers accordingly (Sec. 4.2 in [102], Sec. 4.3 in [106, 107] and Sec. 4.4 in [66]). This chapter extends these publications in terms of depth and breadth of the descriptions, and it classifies and re-evaluates the results in relation to each other.

The remainder of the chapter is structured as follows. In the next section, the controller model as well as a set of mutation and recombination operators, general strategies (elitist vs. non-elitist), fitness functions and the course of evolution in an online and onboard manner are defined. In Sec. 4.2, a preliminary simulation study is described that shows that CA can, in principle, be evolved using the proposed approach. This study further serves as a broad parameter adjustment for the experiments conducted in the subsequent sections. Sec. 4.3 presents another simulation study which includes an extensive testing of several evolutionary parameters with regard to their influence on the evolutionary success. The results show that CA and "Gate Passing" behaviors (see below) can be evolved with a high success rate. Sec. 4.4 presents the results of a real-robot study showing that the approach can be successfully transformed to the robot platform Wanda. Finally, Sec. 4.5 concludes the chapter with a short discussion of the results and an outlook to future work.

4.1 Theoretical Foundations

In this section, the MARB controller model is introduced. Based on this model, a decentralized online-evolutionary framework is defined including operators for mutation and recombination as well as methods for fitness calculation, selection, and, optionally, a decentralized elitist strategy. The evolutionary model is examined theoretically in terms of evolvability (as defined in Sec. 2.1.6 – as opposed to the definition in Chap. 5) and applicability to simulation and real-robot scenarios. The designs of the mutation and recombination operators are justified with respect to their influence on evolvability. Additionally, an encoding of MARBs is presented which can be used to store controllers on real robots or as a transfer format between real robots and simulation. In Chap. 5, the encoding is used as a foundation for the initial universal translator. The encoding is theoretically examined in terms of memory consumption and the applicability of mutation operations to it.

4.1.1 Preliminaries

In this chapter, there is no explicit genotypic representation of controllers, i. e., the mutation and recombination operations are performed directly on the MARB level. However, for the purpose of generality a MARB can be interpreted as its own genotypic representation with the GPM being an identity mapping. Therefore, a MARB will be alternatively referred to as the *genome* or *genotype* of a robot in this chapter. Accordingly, the *genotypic search space* $\mathcal{G}$ as well as the *phenotypic search space* $\mathcal{P}$ are both defined to be the space of all MARBs for now.

Note that, in some sense, there is always a distinction between the descriptive part of robot behavior, which is subject to mutation and recombination, and the actual behavior performed during execution of a controller (i. e., the mapping from a sequence of sensor data to the corresponding sequence of output instructions), which is subject to selection. It might seem plausible to call the descriptive level "genotype", as it is the part which is being altered during evolution, and the behavioral level "phenotype", as this is the actually evaluated part which emerges from a translation process out of the descriptive part. However, it is not common to do so in ER due to the following reasons. Firstly, the behavioral level is rather fuzzy and hard to define in a structured way. Secondly, in classic EC both these terms are used in a descriptive way as there exists no behavioral level there. For compatibility reasons, the according terminology is used here, too. Particularly, the GPM defined in Chap. 5 relies on phenotypes represented by MARBs and genotypes represented by sequences of integers.

From Moore machines to MARBs. The proposed controller model MARB is based on finite *Moore automata* or *Moore machines* as described by Hopcroft et al. [78]. A Moore machine is defined by the tuple

$$A = (Q, \Sigma, \Omega, \delta, \lambda, q_0)$$

where

Q is a set of states (q_0 being the distinguished initial state),

Σ is an input alphabet,

Ω is an output alphabet,

$\delta : Q \times \Sigma \to Q$ is a transition function, and

$\lambda : Q \to \Omega$ is an output function.

The output *op* of a state is interpreted as an operation to be executed by a robot, for example "drive forward for 10 cm" or "turn left by 10 degrees". Therefore, the output alphabet Ω is identified by a set *Op* of all available operations. Information provided by the sensors and the robot's internal state build the input of the automaton and define which transitions to take. There, for simplification the internal state is given by a set of sensor values, too. Thus, the input alphabet consists of all possible combinations of sensor values of a robot.

To be applicable to micro-robotic platforms with low memory and processing capabilities, the model is based on byte values which can be efficiently stored and processed using the `uint8` datatype of the C programming language available on the Jasmine and Wanda robots. As these platforms provide most information such as sensor data in this form, both MARB inputs and outputs are encoded as byte values. Furthermore, the encoding of MARBs presented in Sec. 4.1.3 is defined as a sequence of byte values which is suitable to be easily stored in different storage types given by the robot platforms. Of course, the model can be easily generalized to work with different data types.

Sensor variables. The interface between a robot's perception of the world and the decisions made by its MARB controller is given by a set of sensor variables storing the information relevant for the robot's interaction with the environment. This information is assumed to be given in $n \in \mathbb{N}$ chunks of byte values. Let a set of byte values and a set of positive byte values be denoted, respectively, as

$$B =_{def} \{0, \ldots, 255\} \text{ and } B_+ =_{def} \{1, \ldots, 255\}$$

Let $V =_{def} B^n$ be the space of sensoric information available to the controller. The set of a robot's n sensor variables is denoted by

$$H = \{h_1, \ldots, h_n\} \, (H \cap B = \emptyset)$$

where each sensor variable h_i stores at a time step $t \in \mathbb{R}$ the value $v_i(t)$ from the tuple $v(t) = (v_1(t), \ldots, v_n(t)) \in V$ given by the perceived sensoric information at that time step. For the purpose of readability, the time step t is omitted in the following whenever it is given or irrelevant by context.

In general, the sensor data may originate from *real* or *virtual* sensors. The former can hold any sensoric information or information about the internal state of a robot; the latter can hold information about the translation process from genotype to phenotype, i. e., the GPM. (Virtual sensors are defined in the next chapter as part of a closely related Moore automaton accomplishing this translation process.) In this chapter, seven real sensor variables are utilized which belong to the seven *Infra Red (IR)* sensors of a Jasmine or Wanda robot. More precisely, in simulation each h_i, for $1 \leq i \leq 7$, is associated to the sensor labeled with i in Fig. 4.6b (page 98) and delivers the value v_i at a specific time step. When using real Wanda robots, the sensors are distributed similarly, as shown in Fig. 4.7b (page 99).

Operations. At the actuating side, the operations a controller can cause to be executed are also encoded as byte values; for encoding reasons, the zero value is omitted at this point (cf. Sec. 4.1.3). Let

$$Cmd = \{cmd_1, \ldots, cmd_m\} \subseteq B_+$$

be a set of m *commands*, encoded as positive byte values. A command encodes an action or a series of actions to be executed, given by a method in a robot's program storage. Each command is part of an *operation* which combines the command with up to two parameters that specify how the execution is to be performed. Therefore, theoretically $|B|^3 = 16,581,375$ different operations can be encoded. Some commands ignore the parameters meaning that, in that case, the execution does not depend on the value of one or both parameters. In this chapter, the following commands are utilized for specifying the movement of a robot:

- *Idle* ("keep executing the last command"; both parameters are ignored; encoded as 1),
- *Stop* ("stop executing any commands"; both parameters are ignored; encoded as 2),
- *Move* forward ("let both wheels turn with the same speed in forward direction"; first parameter denotes driving distance or speed as specified below, second is ignored; encoded as 3),
- *TurnLeft* ("let both wheels turn with the same speed in opposite directions leading to a left turn of the robot"; first parameter denotes angle to turn or angular speed as specified below, second is ignored; encoded as 4),
- *TurnRight* ("let both wheels turn with the same speed in opposite directions leading to a right turn of the robot"; first parameter denotes angle to turn or angular speed as specified below, second is ignored; encoded as 5).

Depending on the robot platform or the simulation model used, the concrete execution of these commands may vary. Therefore, it is defined more precisely in the respective sections below.

An operation *op* is defined as a tuple involving a command $cmd \in Cmd$ and two parameters $par_1, par_2 \in B_+$:

$$op = (cmd, par_1, par_2) \in Op =_{def} Cmd \times B_+ \times B_+$$

where
$$Cmd =_{def} \{1, 2, 3, 4, 5\}.$$

In the following, the byte values from the set *Cmd* will be identified with *Idle*, *Stop*, *Move*, *TurnLeft* and *TurnRight*.

MARB Transitions. Transitions of Moore machines are defined for every combination of states and input symbols by the transition function

$$\delta : Q \times \Sigma \to Q$$

For MARBs, it would be infeasible to list all transitions explicitly as the input alphabet consists of all combinations of possible sensor inputs. For $|Q|$ states the length of the list would be $256^7 \cdot |Q|$ in this chapter. Therefore, a language of *conditions* has been developed with the purpose of merging transitions into clusters of "similar" meaning from a behavioral point of view. The language of conditions allows for sensor values to be compared with each other or with constants, and for sub-conditions to be connected by conjunction or disjunction. The set of conditions C over the sensor variables H is defined to be the following set (using the *Extended Backus–Naur Form*):

$$c ::= true \mid false \mid z_1 \triangleleft z_2 \mid (c \circ c),$$

where $z_1, z_2 \in B_+ \cup H$,
$\triangleleft \in \{<, >, \leq, \geq, =, \neq, \approx, \not\approx\}$,
$\circ \in \{AND, OR\}$.

The values *true* and *false* are called *atomic constants*, $z_1 \triangleleft z_2$ is called an *atomic comparison*. Therefore, a condition can be an arbitrary combination of atomic comparisons and atomic constants, connected by *AND* and *OR*. A condition $c \in C$ can be evaluated to *true* or *false* depending on the underlying sensor values $v \in V$, by using the evaluation function given on page 72. For now, a canonical understanding of the meaning of conditions is assumed. Example conditions are:

- *true, false,*
- $h_1 < h_2$ ("sensor 1 has a smaller value than sensor 2"),
- $20 > h_7$ ("sensor 7 has a smaller value than 20"),
- $(h_1 < h_2 \; OR \; h_2 \approx 120)$ ("sensor 1 has a smaller value than sensor 2 or sensor 2 has a value approximately equal to 120, i. e., differing by not more than 5").

Note that for encoding reasons, the constant 0 cannot be part of a condition. However, every atomic comparison containing 0 can be expressed as an equivalent condition without 0, for example, $h_1 > 0$ can be expressed as $h_1 \geq 1$, $h_1 = 0$ as $h_1 < 1$, and $h_1 \geq 0$ as *true*.

Every state q has an associated set $trans^q = \{(c_1, id_1), \ldots, (c_k, id_k)\}$ where the c_i denote a condition, and the id_i encode a destination state which can be visited next if the condition evaluates to *true*; for a state q', this encoding id is given by a byte value and denoted as $id^{q'}$. In the context of MARBs, an element $(c, id^{q'}) \in trans^q$ is called *transition* with source state q, condition c and destination state q'; it is also called *outgoing transition* of q and *incoming transition* of q'. If the condition of a transition evaluates to *true* under given sensor values $v \in V$, the transition is called *active*. If at time step t there is exactly one active transition $(c, id^{q'}) \in trans^q$ for the current state q under the current sensor values $v(t) \in V$, this transition is taken, meaning that the destination state q' is set to be the next state at $t + 1$. Otherwise, the following two special cases have to be considered:

1. If for q none of the outgoing transitions are active, there is an implicit transition $(true, id^{q_0})$ to the initial state q_0 defined; this approach is preferred to the naïve idea of simply defining q itself as the destination state, as experience with an earlier version of the model has shown that this frequently leads to deadlocks.

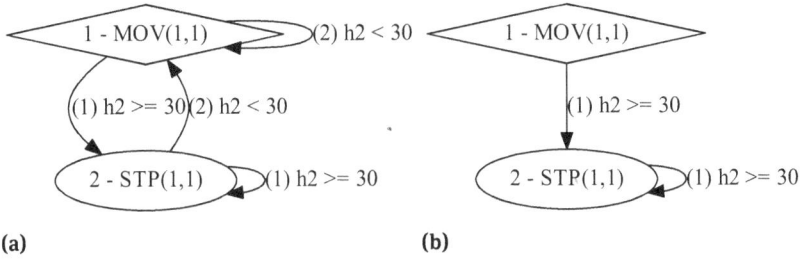

Fig. 4.1. Example MARBs showing an application of implicit transitions. Both transitions pointing to the initial state in (a) can be omitted as depicted in (b). They are implicitly defined for the case that the respective other transition evaluates to false. The automata in (a) and (b) are therefore equivalent.

2. If more than one outgoing transition is active, a prioritization order is defined, and the active transition prioritized highest is taken. The prioritization order used here corresponds to the order of insertion of the transitions during creation of the automaton. The earlier in the automaton building process a transition has been inserted, the higher its priority. Accordingly, a transition's priority is loosely related to the time in the evolutionary process when the transition has first been introduced meaning that older transitions are expected to be preferred over new ones.

A transition $(c, id^{q'}) \in trans^q$ of a *MARB* can be seen as a placeholder for a collection of classic transitions of a Moore machine from state q to state q'. Each combination of sensor values which lets a condition evaluate to *true* is covered by this condition. A *MARB* transition, therefore, can represent a fusion of a large set of classic transitions, greatly improving the compactness of the model. For example, two transitions associated to $h_1 < h_2$ and $h_1 \geq h_2$, respectively, represent together the set of all 256^7 possible outgoing transitions of a state (as does a single transition associated to *true*).

Example MARBs. Fig. 4.1 shows two example MARBs. The initial state q_0 is depicted as a rhombus, all other states are depicted as ellipses. Transitions are depicted as arrows pointing from the source to the destination state. Every state q is labeled by a string of the form "id^q – $cmd(par_1, par_2)$" where $id^q \in B$ is the unique byte value identifying the state as described above, and the rest of the string encodes the operation $(cmd, par_1, par_2) \in Op$ to be executed when the state is visited. Transitions are labeled by a string of the form "(r) c", where $r \in \mathbb{N}$ denotes the ranking of the transition in the insertion order of the source state (i. e., a lower number denotes a higher priority of the transition), and $c \in C$ is the transition's condition. The MARB in Fig. 4.1(a) has two states and a complete definition of transitions, i. e., for every state/input combination there is a transition explicitly defined. Fig. 4.1(b) shows a structurally similar MARB with incomplete definition of transitions. Here, from both states an implicit transition is inserted in the case that the other transition evaluates to *false*. Therefore, both MARBs in the figure are equivalent in terms of their behavior (cf. definition of MARB behavior and semantic equivalence on page 75).

Evaluation of conditions. To evaluate conditions to *true* or *false*, a semantics function has to be defined. As this function depends on the sensor values perceived by a robot, an assignment function of sensor values to sensor variables is given first.

Definition 4.1 (*Assignment function*)

Let $v = (v_1, \ldots, v_n) \in V = B_+^n$ for some $n \in \mathbb{N}$ be (at some time step t) the values of the (real or virtual) sensor variables $h_1, \ldots, h_n$, let $z \in H \cup B_+$. The assignment function A is defined as:

$$A : (H \cup B_+) \times V \to B^+ : A(z, v) = \begin{cases} v_i & \text{if } z = h_i \in H, \\ z & \text{if } z \in B_+ \end{cases}$$

For abbreviation, $A(h_j, v) = v_j$ for $h_j \in H$ is written in the following as:

$$h_j^v = v_j$$

or, if unambiguous, simply as:

$$h_j = v_j$$

The semantics of conditions is defined by a function which takes a condition as input and returns its value, *true* or *false*, as output, based on the current sensor values. Using the assignment function, the semantics can be defined as follows:

Definition 4.2 (*Semantics of conditions*)

Let $c \in C$ be a condition, $v = (v_1, \ldots, v_n) \in V$ the values of the (real or virtual) sensor variables $h_1, \ldots, h_n$, and $A : (H \cup B_+) \times V \to B_+$ the assignment function. The semantics of c is a function

$$S : C \times V \to \{true, false\}$$

with
- $S[\![true, v]\!] = true$
- $S[\![false, v]\!] = false$
- $S[\![z_1 < z_2, v]\!] = \begin{cases} true, & \text{if } A(z_1, v) < A(z_2, v), \\ false & \text{otherwise} \end{cases}$
- $S[\![z_1 > z_2, v]\!] = \begin{cases} true, & \text{if } A(z_1, v) > A(z_2, v), \\ false & \text{otherwise} \end{cases}$
- For a constant $k \in B$ (throughout this thesis set to $k = 5$):

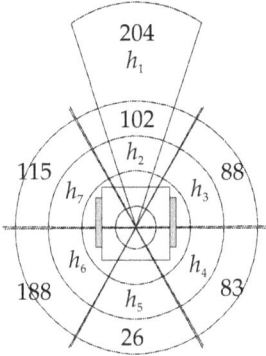

$v = (204, 102, 88, 83, 26, 188, 115)$

Fig. 4.2. Example of the assignment of sensor values to sensor variables. The sensor values $v_1, \ldots, v_n$ perceived by the robot are assigned to the sensor variables $h_1, \ldots, h_n$; they are denoted by the vector $v = (v_1, \ldots, v_n)$. In this chapter, the 7 IR sensors of the Jasmine IIIp and the Wanda robot platform, respectively, have been used exclusively. In Chap. 5, additional virtual sensors are used to make the current state of a GPM translation process perceivable by the translator automaton.

$$S[\![z_1 \approx z_2, v]\!] = \begin{cases} true, & if\ |A(z_1, v) - A(z_2, v)| \le k, \\ false & otherwise \end{cases}$$

- $S[\![z_1 \not\approx z_2, v]\!] = \neg S[\![z_1 \approx z_2, v]\!]$,
- $\ldots$ (likewise for the other comparison operators),
- $S[\![(c_1\ AND\ c_2), v]\!] = S[\![c_1, v]\!] \wedge S[\![c_2, v]\!]$,
- $S[\![(c_1\ OR\ c_2), v]\!] = S[\![c_1, v]\!] \vee S[\![c_2, v]\!]$,

where $z_1, z_2 \in B_+ \cup H$, $c_1, c_2 \in C$, and $\wedge, \vee, \neg$ are the common logic operators.

The following example shows how the evaluation function is used:

Example (Semantics of conditions)

Let $v = (204, 102, 88, 83, 26, 188, 115)$ be the currently perceived sensor values, cf. Fig. 4.2. The sensor variables have the assigned values

$$h_1 = 204, h_2 - 102, h_3 = 88, h_4 = 83, h_5 = 26, h_6 = 188\ and\ h_7 = 115$$

The evaluation function S yields, for example, the following results:
- $S[\![h_1 > h_2, v]\!] = true$.
- $S[\![(true\ AND\ h_2 \approx 134), v]\!] = false$.
- $S[\![true, v]\!] = true$.
- $S[\![false, v]\!] = false$.
- $S[\![((h_5 < 100\ AND\ h_6 > h_5)\ OR\ h_1 \le 95), v]\!] = true$.

Random numbers. As a final preliminary, a function rand(S) is assumed, which returns a random element out of an arbitrary non-empty finite set S based on uniform distribution. In the following, random operations are assumed to be based on uniform distribution if not stated otherwise. In reality, depending on the implementation of the corresponding random number generator, rand may vary in its behavior. For the experiments, the Java built-in random number generator as implemented within the EAS Framework and the random number generator available on the real-robot platforms have been used, respectively. Both these generators depend on an input number called *seed* which determines the list of pseudo-random numbers created by repeatedly invoking the generator. In simulation the seed for each experiment is given itself as a random number by using the random number generator with the current CPU time as seed. Onboard of the robots, the random number generator initializes itself by generating a seed based on a mixture of the currently perceived sensor values.

4.1.2 Definition of the MARB Controller Model

A MARB is defined as a Moore machine given by the previously described constructs. Let $A = (Q, \Sigma, \Omega, \delta, \lambda, q_0)$ a Moore machine. A MARB transition $t = (c, id^q) \in trans^q$ is associated to its respective source state $q \in Q$, therefore, the set of states Q can be used as a basis to define all parts of a MARB. The transition function δ can be defined according to the information given by t about its source state, destination state and condition. The output function λ is naturally associated to the states by assigning an operation to each. The input alphabet $\Sigma = B^n$ contains all possible combinations of sensor values, however, it does not have to be stored explicitly. The output alphabet $\Omega = Cmd \times B_+ \times B_+$ is the set of all operations, and it also does not have to be stored explicitly. In the actual software realization, all parts of a MARB are defined along with the states as this leads to a compact and efficient encoding, and it allows for a rather simple mutation on the encoding level (cf. Sec. 4.1.3). To support this on a formal level, a MARB is defined as follows:

Definition 4.3 (MARB)

Let C be the set of conditions, and Op the set of operations. A MARB is a Moore machine

$$A = (Q, \Sigma, \Omega, \delta, \lambda, q_0)$$

with the following elements:

The set of states $Q \in B_+ \times Op \times \bigcup_{i=1}^{n} (c_i, id_i)$, $c_i \in C$, $id_i \in B_+$, and for each state:

$$q = (id^q, op^q, trans^q) \in Q,$$
$$op^q = (cmd^q, par_1^q, par_2^q) \in Op,$$
$$trans^q = \left\{ \left(c_1^q, id_1^q\right), \left(c_2^q, id_2^q\right), \ldots, \left(c_{n^q-1}^q, id_{n^q-1}^q\right), \left(c_{n^q}^q, id_{n^q}^q\right) \right\}, n^q \in \mathbb{N}.$$

where
- $id^q \in B_+$ *is the state's unique identifier;*
- $op^q \in Op$ *is the operation defining the output of state q;*
- $trans^q$ *is the state's set of outgoing transitions; for any $(c, id) \in trans^q$, c denotes the condition of the transition, and id the identifier of the destination state;*
- $\left(c^q_{n^q}, id^q_{n^q}\right) =_{def} (true, id^{q_0}) \in trans^q$ *denotes q's implicit transition.*

The input alphabet $\Sigma = V = (B_+)^{|H|}$.
The output alphabet $\Omega = Op$.
The transition function $\delta: Q \times \Sigma \to Q$:

$$\delta(q, v) = q', \text{ if and only if } \exists i \in \{1, \ldots, n^q\} : \left(c^q_i, id^q_i\right) \in trans^q \text{ and}$$
$$S\left[\!\left[c^q_i, v\right]\!\right] = true \text{ and } id^q_i = id^{q'} \text{ and}$$
$$\forall \left(c^q_k, id^q_k\right) \in trans^q : S\left[\!\left[c^q_k, v\right]\!\right] = true \Rightarrow k \geq i,$$

for $q, q' \in Q, v \in V$ (the respective perceived sensor values).
The output function $\lambda: Q \to Op: \lambda(q) = op^q$, for $q \in Q$.
The initial state $q_0 \in Q$.

Note that due to the implicit transition in every state's set of outgoing transitions, the transition function is well-defined. Furthermore, as the implicit transition is prioritized lowest of all transitions, the definition matches the description of the two special cases of active transitions. Transitions of a MARB A are also identified by the set

$$T(A) \in (B_+ \times B_+) \times C$$

where

$$\left(\left(id^q, id^{q'}\right), c\right) \in T(A) \Leftrightarrow \exists q, q' \in Q, c \in C : \left(c, id^{q'}\right) \in trans^q$$

A transition $\left(\left(id^q, id^{q'}\right), c\right) \in T(A)$ is given by its source state q, its destination state q' and its condition c.

As stated before at an informal level, the space of all MARBs is called the phenotypic space $\mathcal{P}$ which is equal to the genotypic space $\mathcal{G}$ in this chapter.

MARB behavior. For a MARB A, the sequence of operations $(op(0), op(1), \ldots) \in Op^*$ produced by A as an output when being confronted with a sequence of sensor values $(v(0), v(1), \ldots) \in V^*$ is called the *behavior* of A.

Definition 4.4 (*MARB Behavior*)

Let $A \in \mathcal{P}$ be a MARB; let $t \in \mathbb{N}$ denote a time step, assuming that A performs one state change per time step. Let $v(t) \in V$ be the vector of perceived sensor values at time step t and $q(t)$ the active state of A at time step t.

It holds that $q(0) = q_0$ and $\forall t > 0 : q(t) = \delta\,(q(t-1), v(t-1))$.

MARB behavior up to time step t, BEH^t, is defined as a mapping

$$BEH^t : \mathcal{P} \times V^t \to Op^t :$$

$$BEH^t[\![A, (v(0), \ldots, v(t))]\!] = (\lambda(q(0)), \ldots, \lambda(q(t))).$$

MARB behavior BEH is then defined by letting t approach infinity:

$$BEH = BEH^t, t \to \infty.$$

Let V^∞ denote the set of infinite sequences of sensor value combinations. If for two MARBs A_1, A_2 it holds:

$$\forall(v(0), v(1), \ldots) \in V^\infty : BEH[\![A_1, (v(0), v(1), \ldots)]\!] = BEH[\![A_2, (v(0), v(1), \ldots)]\!],$$

A_1 and A_2 are called *semantically equivalent* which is denoted by:

$$A_1 \equiv A_2.$$

When introducing the mutation operator in Sec. 4.1.4, a distinction will be made between "syntactic" and "semantic" mutations. Both change the structure of a MARB A to a mutated version $M(A)$, but only semantic mutations can also change its behavior. Therefore, for all syntactic mutations it will hold that $A \equiv M(A)$.

4.1.3 Encoding MARBs

The encoding of a MARB A is given by a sequence of byte values $[A] \in B^*$. As stated above, all information required to unambiguously define a MARB is given by the set of states, provided that q_0 is marked as the initial state. Therefore, MARB controllers can be completely encoded by storing the set of states in a list and defining one of its elements as the initial state. The latter is accomplished by setting the initial state to be the first element of the list. Fig. 4.3 shows schematically the encoding up to the level of conditions.

Notations. The following notations and terms are used concerning the encoding of MARBs:

– Byte values are always written with three digits: $000, \ldots, 255$.
– To separate single byte values or larger pieces of code from each other, "|" is used.
– A byte sequence encoding a part of a MARB is denoted by square brackets. This notation is used for all parts of a MARB without further formalization. E. g., $[A]$ is the encoding of MARB A, $[q]$ is the encoding of a state q.
– The set of all correct encodings of MARBs is given by

$$[\mathcal{P}] =_{def} \{[A] \mid A \in \mathcal{P}\}$$

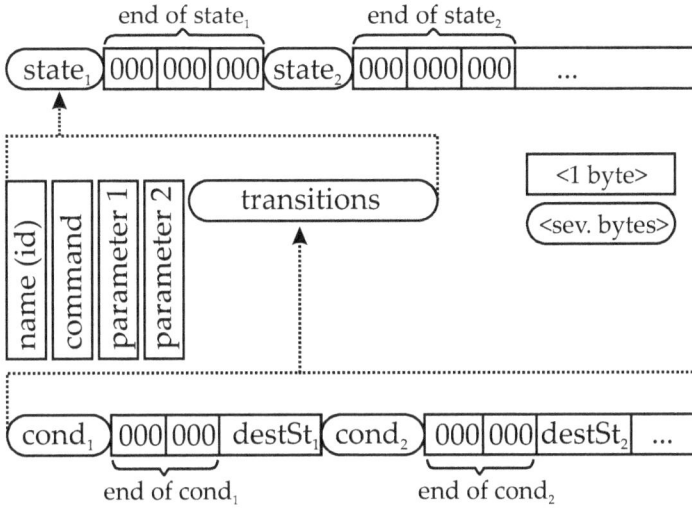

Fig. 4.3. Schematic view of the encoding of a MARB. The set of states Q of a MARB is encoded by storing its elements in a sequential list separated be a sequence of three zero values (top line). For each single state $q \in Q$ the encoding consists of four byte values denoting the state's name id^q and its operation op^q involving a command and two parameters (middle part). Subsequently the set of outgoing transitions $trans^q$ is encoded by storing it in a sequential list. There, each transition consists of an encoding of the condition followed by two zeros and the name of the destination state $id^{q'}$ (bottom line).

- A byte sequence $s \in B^*$ is called a *secure delimiter for a set of sequences* $M \subseteq B^*$, if for any byte sequence $s' \in B^*$ occurring in a sequence $[A] \in [\mathcal{P}]$ it holds that

$$\exists u, v \in B^* : [A] = us'sv \Leftrightarrow s' \in M$$

In other words, in a sequential traversal of a correct MARB encoding from left to right the occurrence of s implies the end of a sequence from M and vice versa. For example, the sequence

$$000 \mid 000 \mid 000$$

can only (and does always) occur after sequences that encode a state and is, therefore, a secure delimiter for these sequences. The second secure delimiter used in the following is the sequence

$$000 \mid 000$$

which occurs only after encodings of conditions.

Encoding the set of states. Let

$$Q = \{q_0, \ldots, q_m\}$$

be the set of states of a MARB A. Q is encoded by storing the encodings $[q_0], \ldots, [q_m]$ of the states sequentially using $000 \mid 000 \mid 000$ as a secure delimiter. The encoding of A is given by:

$$[A] = [q_0] \mid 000 \mid 000 \mid 000 \mid [q_1] \mid 000 \mid 000 \mid 000 \mid \ldots \mid [q_m] \mid 000 \mid 000 \mid 000$$

The order of the states is irrelevant, except for $[q_0]$ which is the first state in the code. Let

$$q = \left(id^q, \left(cmd^q, par_1^q, par_2^q\right), \left\{\left(c_1^q, id_1^q\right), \ldots, \left(c_{nq}^q, id_{nq}^q\right)\right\}\right) \in Q$$

All variables in this statement, except for the c_i, can be encoded by single byte values. The variables $id^q, par_1^q, par_2^q, id_1^q, \ldots, id_{nq}^q$ are already byte values and can be encoded directly using their value in the sequence. To ensure that a sequence of two or three zeros, respectively, can be used as a secure delimiter, the state names, commands and parameters are defined to have non-zero values. The command cmd^q is encoded as follows:

Command	Idle	Stop	Move	TurnLeft	TurnRight
Byte value	001	002	003	004	005

Then, omitting $c_1, \ldots, c_n$, $[q]$ can be written as

$$id^q \mid cmd^q \mid par_1^q \mid par_2^q \mid \underbrace{[c_1^q] \mid 000 \mid 000 \mid id_1^q \mid 001 \mid \ldots \mid [c_{nq}^q] \mid 000 \mid 000 \mid id_{nq}^q \mid 001}_{[trans^q]}$$

The order of transitions in the encoding $[trans^q]$ reflects the transition's priority during evaluation, higher prioritized transitions being encoded first. Two zero values are used as a secure delimiter for the encoding of a condition. An additional byte value has been introduced as a placeholder after the id of the destination states in $[trans^q]$. It is set to a constant value of 001 and ignored throughout this thesis. This placeholder has been used in a related work to store the "age" (i.e., the number of generations since the transition has been altered) and influencing the mutation probability of transitions accordingly [126].

As an example, the MARB A given in Fig. 4.4 can be encoded as follows. The states are defined as (where the implicit transitions "$(true, 1)$" are never taken):

$$q_0 = (1, (Move, 1, 1), \{(h_1 < 90, 1), (h_1 \geq 90, 2), (true, 1)\}), \text{ and}$$
$$q_1 = (2, (Idle, 1, 1), \{(true, 2), (true, 1)\}).$$

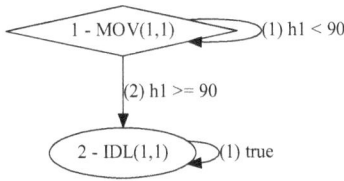

Fig. 4.4. MARB given as an example for the encoding procedure. The MARB's encoding is called [A] in the text.

The encoding [A] can be written as (by leaving out implicit transitions as they are equal for every state and do not have to be stored explicitly):

$$001 \mid 003 \mid 001 \mid 001 \mid [h_1 < 90] \mid 000 \mid 000 \mid 001 \mid 001 \mid$$
$$[h_1 \geq 90] \mid 000 \mid 000 \mid 002 \mid 001 \mid$$
$$000 \mid 000 \mid 000 \mid$$
$$002 \mid 001 \mid 001 \mid 001 \mid [true] \mid 000 \mid 000 \mid 002 \mid 001 \mid$$
$$000 \mid 000 \mid 000.$$

Encoding conditions. The language of conditions contains, besides positive byte values denoting constants or sensor variables, the symbols

$$SY = \{<, >, \leq, \geq, =, \neq, \approx, \not\approx, AND, OR, h, true, false\}$$

Each of these symbols $s \in SY$ is encoded as $[s] = 000 \mid B_s$, i. e., using two byte values the first of which is zero. The second byte value B_s is given in the tables:

s	$<$	$>$	$\leq$	$\geq$	$=$	$\neq$	$\approx$	$\not\approx$
B_s	004	005	006	007	008	009	015	016

s	AND	OR	h	$true$	$false$
B_s	010	011	012	013	014

Conditions are encoded in postfix notation as this allows for a bracket-less encoding and a fast decoding. For example, the following condition in infix-notation

$$((h_1 < h_2 \text{ AND } h_3 \approx h_4) \text{ OR } h_3 < 100)$$

can be converted into postfix-notation as follows:

$$((h_1 h_2 < h_3 h_4 \approx) \text{ AND } h_3 \ 100 <) \text{ OR },$$

and written without brackets:

$$h_1 h_2 < h_3 h_4 \approx \text{AND } h_3 \text{ } 100 < \text{ OR .}$$

The encoding of a condition c is the token-by-token encoding of the bracket-less post-fix notation of c, a token t being a symbol from SY or a byte value within that postfix notation:

$$[t] = \begin{cases} t & \text{if } t \in B_+, \\ 000 \mid B_t & \text{if } t \in SY \end{cases}$$

For example, the condition $c = ((h_1 < h_2 \text{ AND } h_3 \approx h_4) \text{ OR } h_3 < 100)$ is encoded as:

$[c] = [h][1][h][2][<][h][3][h][4][\approx] \text{ [AND] } [h][3] \text{ } [100][<] \text{ [OR]}$

$= 000 \mid 012 \mid 001 \mid 000 \mid 012 \mid 002 \mid 000 \mid 004 \mid 000 \mid 012 \mid 003 \mid 000 \mid 012 \mid$

$004 \mid 000 \mid 015 \mid 000 \mid 010 \mid 000 \mid 012 \mid 003 \mid \text{ } 100 \mid 000 \mid 004 \mid 000 \mid 011$

Finally, the example MARB A is encoded as (underlined parts denote the encoding of conditions):

$001 \mid 003 \mid 001 \mid 001 \mid \underline{000 \mid 012 \mid 001 \mid 090} \mid 000 \mid 004 \mid 000 \mid 000 \mid 001 \mid 001 \mid$

$\underline{000 \mid 012 \mid 001 \mid 090} \mid 000 \mid 007 \mid 000 \mid 000 \mid 002 \mid 001 \mid$

$000 \mid 000 \mid 000 \mid$

$002 \mid 001 \mid 001 \mid 001 \mid \underline{000 \mid 013} \mid 000 \mid 000 \mid 002 \mid 001 \mid$

$000 \mid 000 \mid 000.$

A grammar for the language of correct MARB encodings. The following context-free grammar G_{MARB} defines all byte sequences that are valid encodings of MARBs:

$$G_{MARB} = (\mathfrak{N}, \mathfrak{T}, \mathfrak{P}, S)$$

with
- The set of non-terminals:

$$\mathfrak{N} = \{S, T, C, CMP, OP, TRUE, FALSE, <, >, \leq, \geq, =, \neq, \approx, \not\approx,$$
$$AND, OR, H, SD, CD, B_+\};$$

- The set of terminals $\mathfrak{T} = B$ (the set of all byte values);

– The set of production rules (ϵ denoting the empty word):

$\mathfrak{P} = \{S$ $\rightarrow \epsilon \mid B_+ \, B_+ \, B_+ \, B_+ \, T \, SD \, S,$

T $\rightarrow \epsilon \mid C \, CD \, B_+ \, B_+ \, T,$

C $\rightarrow TRUE \mid FALSE \mid O \, O \, CMP \mid C \, C \, OP,$

O $\rightarrow H \, B_+ \mid B_+,$

CMP $\rightarrow < \mid > \mid \leq \mid \geq \mid = \mid \neq \mid \approx \mid \not\approx,$

OP $\rightarrow AND \mid OR,$

$TRUE$ $\rightarrow 000\,013,$

$FALSE$ $\rightarrow 000\,014,$

$<$ $\rightarrow 000\,004,$

$>$ $\rightarrow 000\,005,$

$\leq$ $\rightarrow 000\,006,$

$\geq$ $\rightarrow 000\,007,$

$=$ $\rightarrow 000\,008,$

$\neq$ $\rightarrow 000\,009,$

$\approx$ $\rightarrow 000\,015,$

$\not\approx$ $\rightarrow 000\,016,$

AND $\rightarrow 000\,010,$

OR $\rightarrow 000\,011,$

H $\rightarrow 000\,012,$

SD $\rightarrow 000\,000\,000,$

CD $\rightarrow 000\,000,$

B_+ $\rightarrow 001 \mid 002 \mid \ldots \mid 255\}.$

The grammar can be used to ensure correctness of MARB encodings (which is done by the MARB simulation FMG within the EAS Framework). Furthermore, it can be used as part of an evolutionary operator to generate MARBs in a purely syntactic way. There, a genotype might encode a traversal through the grammar, starting at S and leading to a terminal word, i. e., a correct MARB encoding. In this case, mutation would have to mutate a genotype in a way that creates a new traversal which, again, leads to a terminal word. However, this approach is not followed here, but left as an outlook to future work (cf. end of this chapter).

Correctness of the encoding. The proposed MARB encoding defines a function $cod :$ $\mathcal{P} \to [\mathcal{P}]$ that maps a MARB A onto a corresponding encoding $cod(A) = [A]$. There also exists an inverse function $dec : [\mathcal{P}] \to \mathcal{P}$ which maps an encoding to a corresponding MARB, such that $dec(cod(A)) = A$ and $cod(dec([A])) = [A]$ for all $A \in P$ and all $[A] \in [P]$. The correctness of these statements is not proven here, but they follow directly from the construction of the code. Moreover, the encoding function cod and the decoding function dec are implemented in EAS in the methods `generateSequence` and `generateFromSequence` in the class `EndlicherAutomat` of package

```
eas.simulation.spatial.sim2D.marbSimulation.endlAutomat.
```

These methods have been tested extensively in various evolutionary runs. They can be reviewed to validate the correctness of the encoding.

Memory consumption of the encoding. As memory is a critical resource on micro-robots, the memory consumption of the proposed encoding is calculated in the following. Let the length of a sequence $S \in B^*$ be denoted by $|S|$. The length of the encoding of a MARB A with the set of states $Q = \{q_1, \ldots, q_m\}$ calculates to

$$|[A]| = \sum_{i=1}^{m} |[q_i]|$$

Let for all $q \in Q$:

$$q = \left(id^q, \left(cmd^q, par_1^q, par_2^q\right), \left\{\left(c_1^q, id_1^q\right), \ldots, \left(c_{n^q}^q, id_{n^q}^q\right)\right\}\right)$$

The memory usage of a state q calculates to

$$|[q]| = 7 + \sum_{j=1}^{n^q} \left(\left|[c_j^q]\right| + 4\right) = 7 + 4n^q + \sum_{j=1}^{n^q} \left|[c_j^q]\right|$$

This sums up the initial four bytes id^q, cmd^q, par_1^q, par_2^q and the three delimiting zeros plus the code for the n^q transitions including four bytes for the two delimiting zeros, the destination state and the placeholder byte (implicit transitions are included in this calculation; replacing n^q with $n^q - 1$ calculates the memory usage without implicit transitions). Thus, the total memory usage of $[A]$ calculates to

$$|[A]| = 7m + \sum_{i=1}^{m} \left(4n^{q_i} + \sum_{j=1}^{n^{q_i}} \left|[c_j^{q_i}]\right|\right)$$

The size of a condition can be specified more precisely. Two different kinds of tokens can occur in a condition that can be further divided according to their memory usage:
- Operators <, >, ..., *AND, OR* (2 bytes).

- Operands:
    - single byte values (1 byte),
    - Boolean values *true* and *false* (2 bytes),
    - sensor variables (3 bytes).

For a condition $c \in C$ let $\#Operators(c)$, and $\#Operands(c)$ denote the number of operators and operands, respectively, that occur in c; let $\#Byte(c)$, $\#Bool(c)$, and $\#H(c)$ be the number of *byte values*, *Boolean values*, and *sensor variables*, respectively, that occur as an operand in c. Then the number of operators of c is one less than the number of operands of c:

$$\#Operators(c) = \#Operands(c) - 1$$

and the length of c is

$$
\begin{aligned}
\|c\| &= \overbrace{\#Byte(c) + 2 \cdot \#Bool(c) + 3 \cdot \#H(c)}^{\text{Operands}} + 2 \cdot \#Operators(c)\\
&= \#Byte(c) + 2 \cdot \#Bool(c) + 3 \cdot \#H(c) + 2 \cdot (\#Byte(c) + \#Bool(c) + \#H(c) - 1)\\
&= 3 \cdot \#Byte(c) + 4 \cdot \#Bool(c) + 5 \cdot \#H(c) - 2.
\end{aligned}
$$

The total memory usage of $[A]$ calculates to

$$
\begin{aligned}
\|[A]\| &= 7m + \sum_{i=1}^{m}\left(4n^{q_i} + \sum_{j=1}^{n^{q_i}}\left(3 \cdot \#Byte\left(c_j^{q_i}\right) + 4 \cdot \#Bool\left(c_j^{q_i}\right) + 5 \cdot \#H\left(c_j^{q_i}\right) - 2\right)\right)\\
&= 7m + \sum_{i=1}^{m}\left(2n^{q_i} + \sum_{j=1}^{n^{q_i}}\left(3 \cdot \#Byte\left(c_j^{q_i}\right) + 4 \cdot \#Bool\left(c_j^{q_i}\right) + 5 \cdot \#H\left(c_j^{q_i}\right)\right)\right) \text{ bytes.}
\end{aligned}
$$

The memory usage is linear in the number of states when assuming (falsely, in general) that the conditions have a constant length. On the other hand, evolved automata tend to have only few states while the conditions are rather complex. For example, for purely reactive behavior there is no need to have more than one state for each of the five commands. Assuming conversely that the number of states is constant, this implies a constant maximum number of transitions, and therefore the memory usage is linear in the length of the conditions in that case. Overall, the memory usage is in $O(m \cdot n^{max} \cdot \|[c^{max}]\|)$ where m is the number of states, $n^{max} =_{def} \max(\{n^q \mid q \in Q\})$ is the maximum number of outgoing transitions of any state, and $\|[c^{max}]\| =_{def} \max(\{\|[c_j^q]\| \mid q \in Q, j \in \{1, \ldots, n^q\}\})$ is the longest encoding of all conditions.

4.1.4 Mutation and Hardening

In contrast to ANNs which originally have been developed as artificial counterparts to reasoning and learning processes in the brains of higher animals, FSMs do not intrinsically provide a set of "natural" operations known to be capable of driving a learning process. Similarly, it is less obvious how mutations in an evolutionary process should be performed due to the discontinuous behavioral effects of operations such as changing source or destination states of transitions, changing the meaning of states as well as inserting or deleting states (whereas changing weights in an ANN or even inserting or deleting neurons are much "smoother" operations). As mutation is a critical evolutionary operator concerning evolvability, it is a major goal of this chapter to propose a mutation operator for MARB controllers which allows for a continuous behavioral improvement during evolution. There, beside providing a set of expectedly smooth alterations on MARBs which still exploit the full search space $\mathcal{P}$, the operator is intended to "harden" parts of a MARB which are highly involved in the currently evolved behavior. Other parts which are less frequently used are left loosely connected and are easier changeable by future mutations. Combined with selection, this mechanism aims at providing an adaptable search space traversal, flexible at first, and more and more congealing at promising areas of the search space in the course of evolution.

Search space structure Using the space of all MARBs as genotypic search space makes this search space "flexible", meaning that there are infinitely many MARBs in $\mathcal{G} = \mathcal{P}$, but if the MARB topology is fixed and the condition size is limited, the according sub-space is finite. Therefore, inserting a state or a transition into a MARB or extending a condition by AND and OR can be seen as a *complexification* of the search space in terms of adding a dimension to it, while removing states or transitions or reducing a condition can be seen as a *simplification* in terms of removing a dimension. Other changes to a MARB do not change the dimensionality of the search space. This view on search space simplification and complexification in MARB evolution is similar to neuro-evolution; there, changing just connection weights of an ANN does not affect the search space dimensionality, but removing a neuron reduces it while adding a neuron yields a complexification. With a flexible search space there is no need to define the search space dimensionality before a run, but the problem of finding an adequate controller complexity is included in the evolutionary process [180]. For this purpose, the mutation operator in this chapter is designed to have the following two properties:

1. Both complexification and simplification are allowed to occur during a run, keeping the search space flexible.
2. During evolution, those parts of the automaton are *hardened* which are expected to be highly involved in "good" behavior.

The first property imitates the idea proposed for neuro-evolution in [180], where, however, only one type of dimensionality changing mutations (complexification vs. simplification) has been allowed to occur in a single run. The second property introduces a new concept which is based on the observation that in the middle of the evolutionary process a MARB contains parts that are highly involved in the currently evolved behavior and others that are rather unimportant for the behavior (for example unconnected states that can never be visited). Such unused parts of a genotype are known to have the potential to improve the evolutionary process as a whole by introducing neutral fitness plateaus [98]. On the other hand, it is important to find a search space complexity that is sufficiently large to contain the solution to the problem while not being too large for an efficient search. Hardening allows for neutral plateaus to exist by still focussing the evolutionary search around promising areas of the search space.

The intention of hardening is to decrease the probability of getting altered by mutation for those parts of a MARB that are expectedly highly involved in the current behavior. At the level of states, this can be accomplished by making states less mutable if they have many incoming transitions as such states are expected to be involved in the behavior. At the level of transitions, the complexity of the associated condition and its "proximity to *true*" (i. e., the number of sensor value combinations that lets the condition evaluate to *true*) can be used as an indicator for the transition's impact on the behavior. Therefore, if a transition's condition is "close to *true*", the mutation operator is unlikely to delete it.

While there exists a rather canonical way of achieving the above-described hardening process within a mutation operator (as described below), the question remains why it is even desirable to harden parts according to their expected involvement in the behavior. Obviously, hardening should affect those parts of a MARB that are involved in "good" behavior in terms of the expected fitness value. This, however, is automatically given when combining a mutation operator, which hardens parts involved in the current behavior, with a selection operator favoring those MARBs which have a high fitness. In the course of evolution, the MARBs existing in a population are expected to encode better behavior than those that have been discarded by selection. There, selection mostly considers those aspects of a behavior that are frequently performed (thus parts of the according MARB that are frequently traversed), meaning that two strongly differing MARBs are similar from a selection perspective as long as the parts being frequently executed are similar. Therefore, hardening parts of MARBs involved in the current behavior in the long term implies hardening parts expectedly involved in good behavior. Of course, in a single robot hardening may well affect parts highly-involved in bad behavior, but if bad parts are hardened, this expectedly leads to the extinction of the according MARB in the long run.

A hardening process which has the above-mentioned properties is accomplished as follows. Mutation is designed to use the number of incoming transitions on the state level, and on the transition level a condition's "proximity to *true*" as well as its complexity, as an indication for how likely it should be for each state and transition to be deleted (this is explained in detail below). Within a single mutation operation, hardening is a completely random operation meaning that there is, other than the expected involvement in the behavior, no control of which part of a MARB gets hardened or which is pushed into a more loosely connected direction. Such a control would involve an extensive analysis of the MARB during evolution which is neither desired nor feasible. Therefore, mutation can make a random part of the automaton more (or less) involved in the behavior by simultaneously making it less (or more) deletable by future mutation operations. Although this procedure is completely random, it is expected to lead to a hardening of those parts which are most involved in desired behavior. For the other parts, mutation remains a solely random change and these parts are not expected to get particularly hardened meaning that they stay more loosely connected than the hardened parts and can get removed within fewer mutations. Using this technique, the complexity of the MARBs in a robot population is expected to adapt more quickly to the complexity of the target behavior.

Construction of the mutation operator. Mutation is defined as a function

$$M^\xi : \mathcal{G} \to \mathcal{G} : A \mapsto M^\xi(A)$$

where ξ denotes a state of a random number generator Ξ. The mutation function generates a MARB $M^\xi(A)$ by altering a MARB $A \in \mathcal{G}$ in a (pseudo-) random way. For better readability, the state of the random number generator ξ is omitted in the following; functions in the scope of mutation and recombination are assumed to depend on it implicitly.

The mutation operator M used in this chapter chooses one out of 11 atomic sub-mutations denoted by functions $M_1, \ldots, M_{11} : \mathcal{G} \to \mathcal{G}$ that each perform a specific type of alteration to a MARB. One single mutation operation is accomplished by selecting one of the 11 sub-mutations randomly according to a probability distribution D, i. e., $M(A) = M_i(A)$ where i is chosen with probability $D(i)$ at every execution of the mutation operator. The sub-mutations can be partitioned into *syntactic* and *semantic* mutations. The former change the structure of a MARB A in a way that does not affect its resulting behavior $BEH[\![A, v^*]\!]$ for any $v^* \in V^\infty$; in that case we have $A \equiv M(A)$. The latter are changing the structure of a MARB in a way that can change the resulting behavior. As single mutations in evolutionary algorithms are supposed to keep the (phenotypic) changes small [201], syntactic mutations can be applied rather unconcernedly as they do not change the behavior of a MARB at all, leaving its resulting fitness expectation unchanged. Semantic mutations, on the other hand, have to be designed in a way that the behavioral changes are kept small. For all sub-mutations M_i of the mutation operator it should hold that $A \approx M_i(A)$, i. e., any MARB should be semantically equal or "similar" to its mutated version. This is accomplished by using few non-syntactic mutations and defining them carefully as explained below.

From a structural point of view the atomic mutations $M_1, \ldots, M_{11}$ can be divided into three general types:

1. insert or remove states;
2. insert or remove transitions;
3. change "labels" of states and transitions – labels are:
    (a) operations on a state level; and
    (b) conditions on a transition level.

Based on this structure, in the following, the 11 atomic sub-mutations are defined as applied to a MARB $A = (Q, \Sigma, \Omega, \delta, \lambda, q_0)$ (note that the elements of A referred to below are actual MARB parts, and do not belong to the encoding $[A]$):

(1) Unconnected states can be inserted arbitrarily without any impact on the behavior; existing states are only removed under conditions where they do not have any impact on the behavior:

M_1 (syntactic): Insert a state without incoming or outgoing transitions, with operation $op = (\mathrm{rand}(Cmd), \mathrm{rand}(\{1, \ldots, 5\}), \mathrm{rand}(\{1, \ldots, 5\}))$.

M_2 (syntactic): Remove a random state $\mathrm{rand}(Q)$ including all its outgoing transitions if it has no incoming transitions; otherwise no alteration is performed. An exception is the initial state q_0 which can only be deleted if it is the only state in the automaton.

M_3 (syntactic): Remove a random state $\mathrm{rand}(Q)$ associated with an *Idle*-operation if all outgoing transitions are associated with *false*. If no such state exists no alteration is performed.

(2) Transitions labeled with *false* can be inserted or removed with no impact on the behavior:

M_4 (syntactic): Insert a transition labeled with *false* between two random states $q_r \leftarrow \mathrm{rand}(Q)$, $q_r' \leftarrow \mathrm{rand}(Q)$ (this slightly hardens the state to which the transition points); if no states exist in the MARB no alteration is performed.

M_5 (syntactic): Remove a random transition $\mathrm{rand}(T(A))$ labeled with *false*; no alteration is performed if no such transition exists.

(3a) The state labels are mutated by randomly choosing a state and adding a small random number, between -5 and 5, to its first parameter. If the parameter would fall below zero by this procedure, which is not allowed by definition, the command of the state gets changed randomly and the parameter is set to its absolute value. Afterward, one is added to the parameter in any case, as zero is not allowed as parameter value, either. As the second parameter is ignored for behavioral commands, it remains unchanged. By this method, the behavioral change is kept preferably small even when changing the command of a state. E. g., for small $X, Y \in B_+$, $(Move, X, .)$ is expected to be similar to $(TurnLeft, Y, .)$. However, the behavioral impact can still be high, which is why this mutation is used with a low probability (cf. Table 4.1):

M_6 (semantic): For a random state's operation $op = (cmd, par_1, par_2) \in Op$, change this state's operation to $(cmd', |par_1 + c| + 1, par_2)$, where

$$c = \mathrm{rand}(\{-5, \ldots, 5\}), cmd' = \begin{cases} cmd, & \text{if } par_1 + c \geq 0 \\ \mathrm{rand}(Cmd) & \text{otherwise} \end{cases}$$

(3b) Transition labels are mutated by selecting a random transition and either performing a simplification or complexification on its condition with no impact on the behavior (M_7, M_8) or by changing a random atomic part of the condition with slight impact on the behavior (M_9). The transition is hardened by complexification and by moving towards *true*. Additionally, single values and sensor variables in a condition can be changed (M_{10}, M_{11}); this, however, can have a drastic impact on the behavior and is used with a low probability (cf. Table 4.1). For the following mutations it holds that no alteration is performed if $T(A) = \emptyset$:

M_7 (syntactic): Let c be the condition of a random transition $\text{rand}(T(A))$. Choose randomly one of the following simplification patterns where $c' \in C$:

$$(c' \text{ AND } true) \rightarrow c',$$
$$(c' \text{ AND } false) \rightarrow false,$$
$$(c' \text{ OR } true) \rightarrow true,$$
$$(c' \text{ OR } false) \rightarrow c'.$$

If at least one subpart c'' of condition c matches the chosen pattern, choose randomly such a matching subpart and perform the simplification on it; otherwise no alteration is performed on A.

M_8 (syntactic): Let c be the condition of a random transition $\text{rand}(T(A))$. Choose randomly one of the following complexification patterns where $c' \in C$:

$$c' \rightarrow (c' \text{ AND } true),$$
$$c' \rightarrow (c' \text{ OR } false).$$

Choose randomly a subpart $c'' \in C$ of c representing a valid condition (such a subpart exists for any condition as the condition itself is such a subpart) and perform the complexification on it.

M_9 (semantic): Let c be the condition of a random transition $\text{rand}(T(A))$. A randomly chosen atomic part $c' \in C$ of c can be moved in small steps closer towards *true* or *false*. Let $a, b \in B_+ \cup H$ where $H \backslash \{a, b\} \neq H$. The following alterations are possible:

$$false \leftrightarrow a = b \leftrightarrow a \approx b \leftrightarrow \begin{matrix} a \leq b \leftrightarrow a < b \\ a \geq b \leftrightarrow a > b \end{matrix} \leftrightarrow a \napprox b \leftrightarrow a \neq b \leftrightarrow true$$

While only one of the above patterns matches each c', the direction (closer to *true* vs. closer to *false*) is chosen randomly. If two alternatives can be chosen at a branch (from $a \approx b$ to the right or from $a \napprox b$ to the left), the alternative is chosen randomly. When mutating *true* and *false* into atomic comparisons, a, b are chosen randomly from H and B_+.

M_{10} (semantic): Let c be the condition of a random transition $\text{rand}(T(A))$. Change a subpart of c which represents a number $i \in B_+$ to

$$\min(\max(i + \text{rand}(\{-5, \dots, 5\}), 1), 255)$$

Table 4.1. Probability distribution for the atomic sub-mutations. The table displays the probability distribution D for choosing one of the atomic sub-mutations during a mutation operation. For any $M_i, i \in \{1, \ldots 11\}$, $D(i)$ is the probability of choosing M_i.

	M_1	M_2	M_3	M_4	M_5	M_6	M_7	M_8	M_9	M_{10}	M_{11}
$D(i)$	4/49	3/49	3/49	7/49	7/49	1/49	4/49	6/49	10/49	3/49	1/49

M_{11} (semantic): Let c be the condition of a random transition $rand(T(A))$. Change a subpart of c which represents a sensor variable $h \in H$ to $rand(H)$.

For all atomic sub-mutations, no alteration is performed if the MARB structure does not allow for this mutation (for example if there is no *false* transition to be deleted when M_5 is applied). In that case the MARB remains unaffected, i.e., $M(A) = A$. The mutation operator M chooses one of the atomic sub-mutations randomly, based on the probability distribution in Tab. 4.1. These probabilities have been determined by reasonable assumptions such as "perform mutations rarely which expectedly have a high impact on the behavior", and by extensive observations in simulation and on real robots, however, without rigorous scientific experimentation. It is certainly possible to further optimize this probability distribution, however, the results presented below suggest that focusing on other aspects seems more promising for improving the evolutionary success.

The mutation operator M is *complete* in the sense that for each two MARBs $A, A' \in \mathcal{G}$, there exists an $n \in \mathbb{N}$ and states of the random number generator $\xi_1, \ldots, \xi_n$ with

$$M^{\xi_1} \circ \cdots \circ M^{\xi_n}(A) = A'$$

This is apparent by the following proof sketch. The empty MARB can be generated from every MARB A by
1. reducing all conditions to *false* using M_7 and M_9;
2. deleting all transitions using M_5;
3. deleting all states (which now cannot have any incoming transitions) using M_2.

From the achieved empty MARB, every topology can be derived by using M_1 and M_4. At this point, all conditions are *false*. Using $M_7, \ldots, M_9$, every condition can be produced from *false* by complexification. As all state labels already are arbitrary, this approach can prove that every automaton A' can be derived from every automaton A by a repeated application of M, i.e., M is complete.

Furthermore, the mutation operator is designed to be *smooth* in the sense that a single mutation causes only a small behavioral change. This is accomplished by a careful composition and use of semantic mutations. The property of being "smooth" is hard to objectify, however, there is a line of argument (left out here for space reasons) indicating that M is indeed expected to cause only small behavioral changes to a MARB [100].

4.1.5 Selection and Recombination

In the presented approach, there is no central computer collecting and providing global information about the evolving robot population. Therefore, selection and recombination cannot be implemented as global operators, but have to be accomplished in a decentralized way. Throughout this thesis, a local selection operator is used that can be seen as a decentralized version of tournament selection [54]. Selection is performed whenever two (or several, depending on the setting) robots happen to "come close to each other"; these robots are called a *mating tournament*. More precisely, any sub-population P of the evolving robot population is a mating tournament (and each robot from P is then called *parent*) if $|P| = p$ for a predefined number $p \in \mathbb{N}$, and the distance of any robot from P to at least one other robot from P deceeds a threshold distance ds. Any time a mating tournament occurs during evolution, a *reproduction* is performed on it, i. e., a set of p offspring MARBs is derived from the p parental MARBs by possibly applying recombination, and the resulting MARBs are redistributed among the parents by replacing their former controllers. This process is both local and applicable to real robot populations by setting the threshold distance ds smaller than the IR communication radius of the robots. Then, robots can detect if they are within reproduction range of any other robot by constantly sending and receiving broadcast signals, and afterward a secure communication channel can be established to exchange MARBs and other information. In an actual implementation on real robots, there never have to be more than two robots communicating simultaneously even if $p > 2$, as the process can be sequentialized cf. Sec 4.4.

This completely decentralized reproduction model is used in real-robot experiments as well as in simulation experiments described in Chap. 6. In this and the subsequent chapter, however, a closely related, but more synchronized procedure has been used to establish mating tournaments in simulation. There, a selection interval has been introduced which defines points in time when all robots at once induce a mating procedure. The mating tournaments are established by every robot selecting the $p - 1$ robots closest to itself for offspring production. This approach, while introducing a centralized component which is not directly compatible with the EE approach, is used for simulation runs only, as it is more comprehensive and controllable in experiments than the environment-driven real-robot version of reproduction. However, the synchronized reproduction model is close to a completely decentralized approach making the results transferable to real robots as shown below.

Within a mating tournament, both on real robots and in both versions of selection in simulation, a randomized fitness-proportional selection is performed which selects one robot to provide the topology of the offspring MARBs for all robots in the tournament. In a recombination-free selection the selected robot also provides for the labels, i. e., state operations and conditions, of the offspring MARBs meaning that all robots in the tournament receive an exact copy of the selected robot's MARB. When recombination is used, the offspring MARBs can be generated by a more complex procedure from the parents.

The recombination operator *Cross* studied in this chapter does not affect the topology given by the selected parent's MARB. Rather, it applies the same fitness-proportional selection scheme described above in a second stage to the state and transition labels of the selected MARB topology. There, the value of each operation or condition of this MARB is given by one of the parental robots which is selected for the according state or transition in the second selection stage. As the topologies of the MARBs within the tournament can differ, it can happen that a state or transition of the selected MARB is not present in one or several of the other MARBs in the tournament. In that case the state or transition is labeled by the operation or condition, respectively, given by the original MARB selected in the first stage.

The recombination operator $Cross : (\mathcal{G} \times \mathbb{Z})^p \rightarrow \mathcal{G}^p$ is a mapping from a constant number p of parental genotypes $G_1, \ldots, G_p \in \mathcal{G}$, i. e., MARBs, with corresponding fitness values $f_1, \ldots, f_p \in \mathbb{Z}$ onto p offspring genotypes. In Alg. 4.1, the algorithm is shown for producing one offspring MARB from p parental MARBs with according fitness values, when the formerly selected MARB $G_{sel} \in \{G_1, \ldots, G_p\}$ is given. A set of p offspring MARBs can be produced by repeating the algorithm p times.

Experience shows that most of the MARBs of a single generation of robots during evolution tend to have nearly identical topologies which makes the use of the recombination operator *Cross* reasonable. Otherwise, if extreme cases of completely differing topologies would frequently occur, this operator would basically be diminished to the recombination-free version described above. The recombination operator *Cross* is studied in Sec. 4.3 where it is compared to the simple reproduction method without recombination.

4.1.6 Fitness calculation

Two fitness functions are studied in this chapter, one designed for evolving the task of CA, the other for a more complex task called *Gate Passing (GP)* which is a combination of CA with an additional subtask of passing a gate in the middle of the field as frequently as possible, cf. description below and Fig. 4.5. As a robot's behavior depends on interactions with a complex environment, there is no direct mapping from a genotype to a fitness value. Therefore, a rather fuzzy mapping has to be accomplished at the behavioral level by observation and an estimation of an accurate fitness value cf. Chap. 2 and [86, 87, 139, 146]. As the behavior has to be observed over time before an adequate fitness value can be calculated, this inherently leads to a delayed fitness, not evaluating the current behavior, but the behavior performed in a more or less recent

input $: \left((G_1, f_1), \ldots, (G_p, f_p)\right) \in (\mathcal{G} \times \mathbb{Z})^p; G_{sel} \in \{G_1, \ldots, G_p\}.$
output: $G \in \mathcal{G}.$

Let $G := G_{sel} = (Q, \Sigma, \Omega, \delta, \lambda, q_0);$

for *all states* $q \in Q$ **do**
    if *a state* q' *exists in all MARBs in* $\{G_1, \ldots, G_p\}$ *with* $id^q = id^{q'}$ **then**
        Select a MARB $G' \in \{G_1, \ldots, G_p\}$ by fitness proportional distribution*;
        Let q' be the (only) state in G' with $id^q = id^{q'}$;
        Set $Op^q := Op^{q'}$;
    end
end

for *all transitions* $t = ((id^{q_1}, id^{q_2}), c) \in T(G)$ **do**
    if *a transition* $t' = \left((id^{q'_1}, id^{q'_2}), \ .\ \right)$ *exists in all MARBs in* $\{G_1, \ldots, G_p\}$ *with*
    $id^{q_1} = id^{q'_1}$ *and* $id^{q_2} = id^{q'_2}$ **then**
        Select a MARB $G' \in \{G_1, \ldots, G_p\}$ by fitness proportional distribution*;
        Let $t' = \left((id^{q'_1}, id^{q'_2}), c'\right)$ be the (only) transition in G' with $id^{q_1} = id^{q'_1}$
        and $id^{q_2} = id^{q'_2}$;
        Set $c := c'$;
    end
end

return G;

* The probability to be selected is fitness proportional for positive fitness values only; negative values are treated as 0. If all values are 0, the distribution is defined to be uniform.

Algorithm 4.1: Computing an offspring by the recombination operator *Cross*.

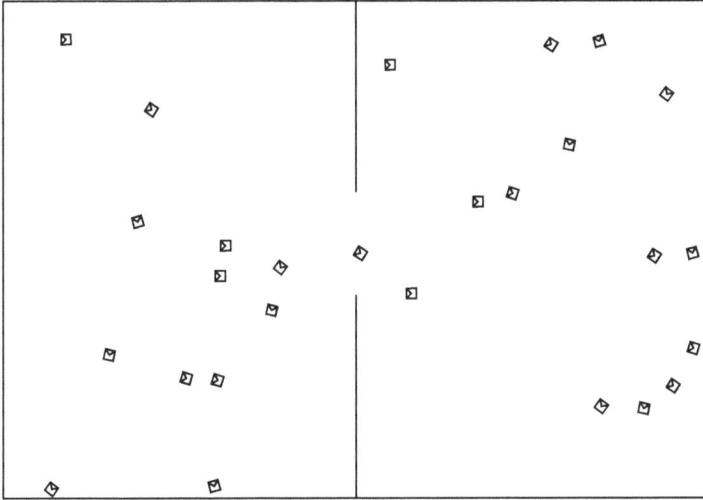

Fig. 4.5. Environment used for the evolution of GP behavior. The robots are supposed to cross the gate as frequently as possible by additionally avoiding collisions. Several robots are drawn to scale.

past. Furthermore, due to the decentralized approach, fitness cannot be calculated by some global mechanism, but has to be determined based on the sensor values and the inner state of a single robot.

Here, a robot's fitness is calculated as the weighted sum of a series of *fitness snapshots* where a snapshot is an integer calculated from the sensor values received by the robot at a certain point in time. Snapshots are calculated at a constant time interval for every robot separately. In simulation this means that all robots calculate snapshots at the same time. On real robots the calculation is not synchronized which leads to a dispersion within the population during an evolutionary run. This has to be considered when transferring the approach from simulation to real robots (cf. Sec. 4.4).

A fitness snapshot provides static information about the quality of a robot's behavior. However, a single snapshot usually cannot be considered an accurate fitness estimation as complex behavior can involve passing through unfavorable situations during the achievement of a goal, and undesired behavior can involve achieving high fitness snapshots by accident. Furthermore, every change of a robot's controller (i. e., mutation or recombination) leads to a diminished accuracy of the earlier fitness snapshots. For example, for a formerly bad automaton which changed to a good one, the old observation of the behavior should lose its influence in favor of the new one. Therefore, the fitness snapshots should be weighted in a way that considers a robot's behavior over a certain adequate time span by still not overrating older values. To achieve this, the fitness snapshots of old behaviors are *evaporating*, i. e., their weight decreases exponentially with the time elapsed. The fitness calculation is divided into two parts which are constantly repeated during a run:

1. *Fitness snapshot* calculation: A rating $snap_X^R(t)$ of the currently observed situation, perceived through the sensors and the inner state of a robot R at time step t, is added to the fitness f_X^R of the robot:

$$f_X^R := f_X^R + snap_X^R(t) \; (X \in \{CA, GP\}, t \in \mathbb{R})$$

This procedure is performed at discrete time steps $t_0, t_1, \ldots$ ($t_{i+1} - t_i = t_{snap} \in \mathbb{N}$ being constant for $i > 0$). After a while, as more and more snapshots are summed up, the expected error caused by differing environmental circumstances of different robots is reduced and f_X^R is expected to adequately reflect the behavior X.

2. *Fitness evaporation* calculation: At time steps $t_0^E, t_1^E, \ldots$ ($t_{i+1}^E - t_i^E = t_{evap} \in \mathbb{N}$ being constant for $i > 0$), a robot's fitness f_X^R is divided by a constant E: $f_X^R := f_X^R / E$. This evaporation procedure accomplishes the exponential decrease in snapshot weight over time.

More formally, the fitness f_X^R for behavior X of a robot R at a time step $t \in \mathbb{N}$ is given by:

$$f_X^R = \sum_{i=0}^{t} s_X^R(i) \cdot c_i$$

where

$$s_X^R(i) = \begin{cases} snap_X^R(i), & \text{if } i \bmod t_{snap} = 0 \\ 0 & \text{otherwise} \end{cases}$$

and

$$c_i = E^{-\lfloor t/t_{evap} \rfloor + \lfloor i/t_{evap} \rfloor}$$

Both parts of the fitness calculation are performed separately in simulation to allow for a flexible adjustment of the snapshot time t_{snap} and the evaporation time t_{evap}. On real robots, however, both snapshot time and evaporation time are not synchronized. As a consequence, different robots may be at different stages of snapshot and evaporation calculation at the same time during an experiment. To avoid unfair selection due to one robot having performed evaporation recently and another a longer time ago, evaporation time and snapshot time are set to the same constant performing evaporation right after the snapshot calculation.

The fitness snapshot for CA behavior. For the CA fitness snapshot, two desired properties are considered:

1. a robot is supposed to drive around covering a preferably long distance per time, and
2. a robot should collide as rarely as possible.

The snapshot for CA is calculated by Alg. 4.2. Note that all input required for the algorithm can be provided by the robot itself.

---

input : Current operation $op \in Op$ of a robot R at a time step t; number of
collisions $|Coll|$ of R since last snapshot before t.
output: Fitness snapshot $snap_{CA}^R(t)$.

let $snap := 0$;
if $op = (Move, X, .)$ *for some* $X \in B$ **then**
  | $snap := snap + 1$;
end

$snap := snap - 3 \cdot |Coll|$;

return $snap$;

Algorithm 4.2: Computation of a fitness snapshot $snap_{CA}^R$ for CA behavior.

The fitness snapshot for GP behavior. The GP fitness snapshot has the additional property that a reward is given when the robot passes a gate in the middle of the field, cf. Fig. 4.5. It is calculated by Alg. 4.3 which is based on Alg. 4.2. The information required for this algorithm as an input can, in principle, be provided by the robot, too. However, it is more complicated to notice a gate passage solely through the robot's sensors (cf. discussion in Sec. 4.4).

---

input : Current operation $op \in Op$ of a robot R at a time step t; number of
collisions $|Coll|$ of R since last snapshot before t; Boolean value $Gate$
indicating if the gate was passed since the last snapshot before t.
output: Fitness snapshot $snap_{GP}^R(t)$ for R at time step t.

let $snap := snap_{CA}^R(op, |Coll|)$;

if $Gate$ **then**
  | $snap := snap + 10$;
end

return $snap$;

Algorithm 4.3: Computation of a fitness snapshot $snap_{GP}^R$ for GP behavior.

4.1.7 The Memory Genome: a Decentralized Elitist Strategy

In ER and, particularly, in an online-evolutionary approach such as EE, where finding and exploiting good behaviors as quickly as possible is essential, the loss of already found good behaviors might be unacceptable for practical applications. The *memory genome* is an elitist strategy for decentralized evolution, as proposed in earlier work [106], to avoid the loss of already available good behavior in a population. It provides a storing mechanism which saves the best-rated genome found up to a certain point in evolution by a single robot. The memory genome's function is to preserve the good genotypes over the course of evolution from the destructive effects of mutation. For this purpose every robot R has a reserved memory capacity to store the MARB MG^R with the highest fitness obtained so far at a time t_{best} including its fitness value at that time $f_X^R(t_{best})$. The stored MARB is updated after every change of the current MARB controller of the robot. If the current MARB's fitness $f_X^R(t)$ falls below a predefined threshold, $f_X^R(t) < th_{MG}$, and if $f_X^R(t) < f_X^R(t_{best})$, the current MARB is replaced by the stored MARB MG^R. However, as the fitness calculation is delayed and noisy, this test is not performed continuously, but in a constant time interval which is a parameter to the evolutionary setting.

4.1.8 Fitness Adjustment after Mutation, Recombination and Reactivation of the Memory Genome

It seems reasonable that a fitness adjustment after mutation and recombination should not ignore all fitness calculations made so far, as the new automaton is expected to be similar to the non-mutated version which has already been evaluated (cf. discussion on "smoothness" of the mutation operator above). Therefore, after mutation, a robot's fitness value is left unchanged, and the adjustment to the new automaton is achieved by the combination of fitness snapshots and evaporation outlined above. After reproduction, the fitnesses of all robots in a tournament are set to the fitness of the robot which has won the tournament (independently of whether recombination is used or not). After reactivation of the memory genome, the fitness is set to the former fitness $f_X^R(t_{best})$ of the stored MARB.

4.1.9 The Robot Platforms

In principle, the MARB controller model is applicable to various robot platforms, solely requiring the existence of a mapping of sensor data onto numerical values, and of numerical values onto actuator control. So far, the MARB model has been implemented and tested on real Jasmine IIIp and Wanda robots in the scope of this thesis, and on a robotic exoskeleton for a medical study performed by researchers from South

Korea and the ETH Zürich [88]. More precisely, the simulation experiments in this thesis have been performed using a model of the Jasmine IIIp robot platform while the Wanda platform has been used for real-robot experiments. Jasmine IIIp has additionally been tested in preliminary real-robot experiments [100, 105]. The Jasmine IIIp and Wanda robot platforms as well as the implementation of Jasmine IIIp in simulation are described in the following.

The Jasmine IIIp platform. Jasmine IIIp is a micro-robotic platform with an approximately cubic body sized $26{\times}26{\times}26$ mm^3, cf. Fig. 4.6a. Jasmine IIIp robots can process simple motoric commands such as to drive forward or backward with a specific speed up to about 1 m per second, or to turn left or right. The robots have seven IR sensors placed around their bodies as depicted in Fig. 4.6b, yielding values from 0 to 255 in order to measure distances to obstacles. Six of the sensors have an opening angle of 60 degrees covering every direction on the movement plane of a robot; the seventh sensor has a smaller opening angle of about 20 degrees and faces to the front with the purpose of detecting more distant obstacles. The IR sensors consist of separate IR transmitters and receivers which can be additionally used to exchange messages between robots. A more technical description of the robot including construction details of its two main boards, motors and sensors, as well as photographies and several experiment recordings can be obtained at www.swarmrobot.org.

In simulation, a robot moves 4 mm (mapped on real world dimensions) straight forward or turns left or right by an angle of 10 degrees per simulation tick. A crash with an obstacle (i. e., the attempt to move at a position that is already occupied by a wall or another robot) is simulated by placing the robot at a random free space within a 4 mm radius from the last position before the collision, and turning it by a random angle, if any of that is possible without a new collision; otherwise only one or none of these actions is performed – the latter, however, happens rarely in very crowded situations only, and gets usually resolved within few further simulation ticks. Deadlocks have never been observed using this collision simulation. The return value of a sensor is calculated in simulation by the function $d(x) = \left\lfloor 255 \cdot 51^{(r-(x \cdot a \cdot b))/150} \right\rfloor$, where x is the distance from the middle of the robot to the closest object in the range of the sensor (in mm) and r is the distance of the sensor from the center of the robot. The parameter a is 1 if the obstacle is a wall, and 2 if it is another robot, simulating different reflection properties of the obstacles' surfaces; b is 1 for the sensors 2 to 7, and 0.75 for sensor 1, simulating that the sensor facing to the front can sense obstacles at greater distances. This sensor calculation function has been developed in preliminary tests with several real Jasmine IIIp robots [100].

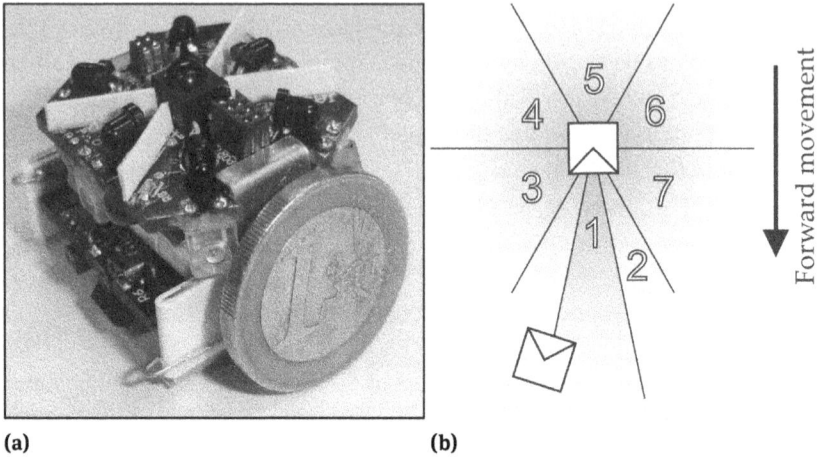

(a)   (b)

Fig. 4.6. Photography and sketch of Jasmine IIIp robots in reality and simulation. (a) Photography of a real Jasmine IIIp robot facing to the bottom-left of the picture. (b) Two simulated Jasmine IIIp robots facing toward each other. IR sensors for distance measurement and obstacle detection are suggested around the above robot; sensors 2 to 7 are using light sources with opening angles of 60 degrees to allow for a gapless detection of obstacles in every direction. Sensor 1 has an angle of 20 degrees to allow detection of more distant obstacles in the front.

The Wanda platform. Wanda is a tube-shaped robot with a diameter of 51 mm and a height of approximately 45 mm, cf. Fig. 4.7a. Its hardware consists of five modular layers as depicted in Fig. 4.8. Each of these layers consists of a *Printed Circuit Board (PCB)* equipped with various elements such as sensors, actuators, memory, processors or batteries. The boards operate independently and are connected by I^2C, UART and SPI buses. The CPU is a Cortex-M3 from Texas Instruments with ARM architecture, clocked at 50 MHz and equipped with 64 kB of RAM and 256 kB of flash memory. It hosts the self-manufactured operating system *WandaOS* derived from Symbricator-RTOS [187], which in turn is based on FreeRTOS [159].

Two *Direct Current (DC)* motors are located on the MB providing a differential control of the wheels. They can induce a forward movement with up to 30 cm per second, or a steering motion when set to different speeds. The inside of the wheels is equipped with a reflective stripe pattern used to measure the actual rotation speed of each wheel and constantly adapting the speed to the desired value (such a feature is not available on Jasmine IIIp).

Six IR sensors (labeled 2 to 7) are attached to the CTB, positioned in a circular way around the middle of the robot, cf. Fig. 4.7b, similar to the arrangement of the Jasmine IIIp robot's IR sensors. An additional IR sensor (labeled 1) is placed at the front of the BB providing a high-precision measurement of obstacles within very short distance ranges (in contrast to the long-distance front sensor of Jasmine IIIp). This dif-

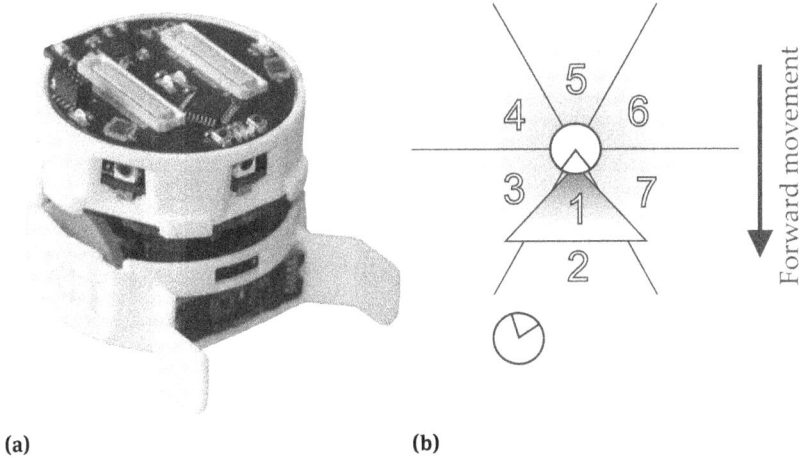

(a)   (b)

Fig. 4.7. Photography and sketch of Wanda robots in reality and simulation. (a) Photography of a Wanda robot including a white plastic shell for IR reflection improvement, facing to the bottom-right of the picture. (b) Sketch of a simulated Wanda robot with the arrangement of the IR sensors $1, \ldots, 7$ suggested around its corpus. The sensors $2 \ldots 7$ work equivalently to the sensors of a Jasmine IIIp robot. Conversely, sensor 1 is a "touch" sensor for short-distance rather than long-distance obstacle detection.

ference between the platforms has to be considered when comparing simulation with real-robot experiments. The six regular sensors can be used to measure distances to obstacles in a range of approximately 35 cm; additionally, as with Jasmine IIIp's IR sensors, a reliable robot-to-robot communication can be established within this range. For communication, an interrupt-based *Time Division Multiple Access (TDMA)* protocol [79] with frames of 100 ms length, featuring ten time slots per frame, has been used. To avoid packet collisions a simple self-synchronizing algorithm has been used establishing a bandwidth of approximately 20 bytes/s. Additionally, ZigBee, a wireless network standard based on IEEE 802.15.4 and suitable for distances up to 100 m is available. By utilizing one Wanda robot as a *host controller* the other robots can exchange information on a de-facto global level as the arena used as an experimentation environment is far smaller than 100 m in each dimension.

The energy required for autonomous operation is obtained from two rechargeable lithium-polymer batteries, one located on the AUXB the other on the BB. They can provide an autonomy time of up to two and a half hours. The AUXB provides a microSD card slot which is used to store MARB history data as well as crash recovery data during evolution. Furthermore, Wanda includes several color LEDs, three IR floor sensors, a 3D accelerometer, an RGB color sensor board at the front of the robot and various other features. To increase the robots' visibility for their own IR sensors, a white plastic shell

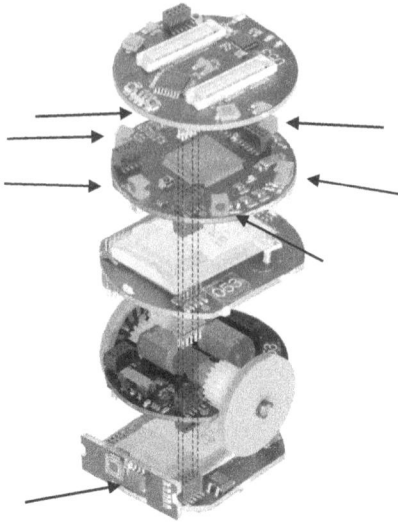

Fig. 4.8. Hardware layers of a Wanda robot. From top to bottom the layers are: *ODeM-RF Board (OB)*, *Cortex Controller Board (CTB)*, *Auxiliary Battery Board (AUXB)*, *Motor Board (MB)*, *Bottom Board (BB)*. Arrows point to the six IR sensors on the CTB and the touch sensor on the BB. Two batteries are located on the AUXB and the BB. Dashed lines indicate the interconnection buses between the boards.

has been put around every robot during the evolution experiments. More details on Wanda are given in a descriptive paper published in 2010 [96].

4.2 Preliminary Parameter Adjustment using the Example of Collision Avoidance

In this section, a set of experiments is described that has been performed to adjust some of the most important evolutionary parameters, namely

- the mutation interval t_{mut},
- the reproduction interval t_{rec},
- the fitness snapshot interval t_{snap}, and
- the fitness evaporation interval t_{evap}.

In the following, first these parameters are described in detail and the remaining parameters left to the model are specified; these specifications are valid for this section only. Subsequently the method of experimentation is given and the results are discussed.

4.2.1 Specification of Evolutionary Parameters

Reproduction. As described above, on real robots reproduction is performed every time a subpopulation of robots forms a tournament, i. e., its members come closer to each other than a given threshold. In simulation, a simplified version of reproduction is performed by using a constant time interval t_{rec} which defines points in time $t \in \{c \cdot t_{rec} \mid c \in \mathbb{N}\}$ at which a robot reproduces, regardless of how far its closest neighbors are. During reproduction, every robot chooses the robot spatially closest to itself (i. e., the number of parents is $p = 2$) to build a tournament and perform the mating procedure without recombination described above. In this section, to simulate the real-robot reproduction process more accurately, for every robot a separate timer is running in a separate thread and the interval t_{rec} is given in real-world seconds. Due to delayed requests and undefined processes running on the simulating computer, the robots' individual times drift apart during the experiment. Therefore, the robots do not reproduce, mutate, calculate fitness, etc. synchronously, but by using only approximately the same time intervals

Mutation. Mutation is performed by applying the operator M defined above at every point in time $t \in \{c \cdot t_{mut} \mid c \in \mathbb{N}\}$ to every robot. The parameter t_{mut} is also given for each robot individually in real-world seconds.

Fitness calculation. The experiments in this section are performed by using the CA behavior as a benchmark. Therefore, the fitness function is calculated by using the fitness snapshot $snap_{CA}^R(t)$ which is added to a robot's fitness value at time steps $t \in \{c \cdot t_{snap} \mid c \in \mathbb{N}\}$. Evaporation is performed every t_{evap} time units, dividing the fitness value of a robot by $E = 2$. The parameters t_{snap} and t_{evap} are given individually for each robot in seconds, too.

4.2.2 Method of Experimentation

Experiments have been performed to adjust the four parameters mutation interval t_{mut}, reproduction interval t_{rec}, fitness snapshot interval t_{snap} and fitness evaporation interval t_{evap}. CA has been chosen as a target behavior, as it is a simple and analyzable behavior which is frequently used in the literature. Beforehand, it has not been clear if CA is evolvable using a MARB-based controller model, therefore, the main goal of this study has been to find any parameter combination which is capable of establishing a successful evolutionary run. The parameter values to be tested have been chosen by intuition and by experience with runs performed on real Jasmine IIIp robots [100, 105].

For every studied parameter, three different values have been chosen, cf. Tab. 4.2, which have been combined in every of the 81 possible ways using a fully factorial experimental design. Every parameter combination has been studied in 10 simulation runs using different random seeds leading to 810 simulation runs overall. Every run has been performed in a rectangle environment sized 600 mm × 800 mm (in

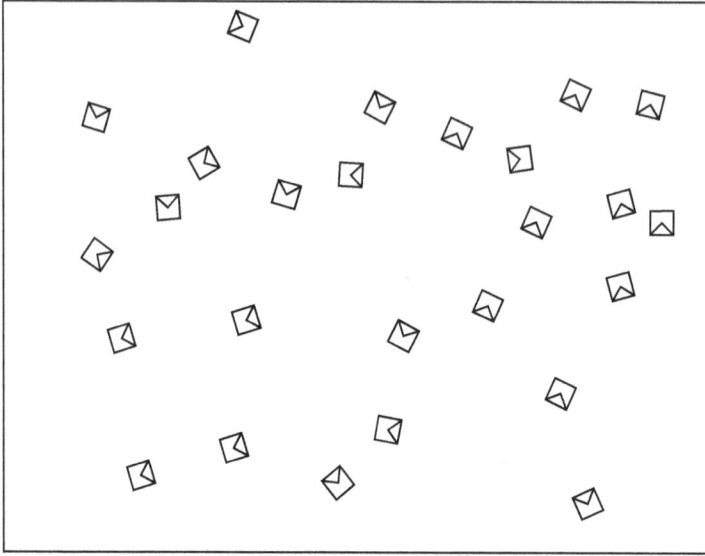

Fig. 4.9. Environment for CA. The environment is sized 600 mm × 800 mm and surrounded by walls. 26 robots are placed at random positions before a run.

scale with the size of the robots) and surrounded by walls, cf. Fig. 4.9. 26 Jasmine IIIp robots have been placed randomly in position and angle at the beginning of every run. Initially each robot's MARB controller has been completely empty, i. e., without any states. Simulation has been performed for about 2,000,000 simulation steps which roughly corresponds to 375 min of real-world time when relating the distance driven by a simulated and a real robot to each other (assuming they constantly repeat the *Move* command). There have been no additional obstacles in the arena meaning that the four surrounding walls and the robots themselves are the only obstacles.

Due to a high number of evolving robots (in total 21,060), and even more MARBs generated during the process, it is obviously not possible to observe all resulting behaviors explicitly. On the other hand, comparing only fitness values to each other is insufficient as fitness calculation itself has been subject to evaluation (by using dif-

Table 4.2. Values of the evaluated parameters used in the experiments. The values have been tested in all possible combinations using a fully factorial experimental setup.

	Mutation Int. t_{mut}	Reproduction Int. t_{rec}	Snapshot Int. t_{snap}	Evaporation Int. t_{evap}
1.	5,000 ms	1,000 ms	250 ms	10,000 ms
2.	10,000 ms	2,000 ms	500 ms	20,000 ms
3.	20,000 ms	10,000 ms	1,000 ms	30,000 ms

ferent values for t_{snap} and t_{evap}). Therefore, a more coarse measure based on fitness values has been introduced to classify evolved behaviors as successful or unsuccessful. The measure evaluates robots by the property of having achieved a positive fitness in the last generation of evolution; robots which fulfill this property are called *successful robots*. An evolutionary run that includes at least one successful robot is called a *successful run*.

Using the method of fitness calculation described earlier, it seems reasonable to define success around the zero fitness barrier, as robots which show good CA behavior are expected to achieve a series of mostly positive fitness snapshots whereas robots which do not show CA behavior either gain zero fitness for not doing anything or achieve a series of mostly negative fitness snapshots. In any case, CA-performing robots are expected to be dragged away from zero fitness in a positive direction while the others are not. To further justify this definition of success, MARBs of successful robots and MARBs of unsuccessful robots (i. e., those with a negative or zero fitness) throughout all experiments have been analyzed concerning their capability of performing CA behavior. The property of having a reachable *Move* state has been used to detect this capability, as MARBs without a reachable *Move* state are clearly not able to perform CA behavior while MARBs which involve a reachable *Move* state may be. The analysis revealed that 91 % of the MARBs of successful robots had a reachable *Move* state, but only 28 % of unsuccessful robots did. Moreover, the behavior of a fairly large set of random samples from both groups has been observed manually showing that it is reasonable to use zero fitness as a barrier between successful and unsuccessful CA behavior. In subsequent studies it has additionally proven useful to generalize the measure to GP behavior, cf. Sec. 4.3.

4.2.3 Evaluation and Discussion

Fig. 4.10 shows the absolute number of successful runs distributed among the 81 different parameter combinations. The highest value (5 out of 10 successful runs) has been achieved by the parameter combination $t_{snap} = 500$ ms, $t_{evap} = 30{,}000$ ms, $t_{rec} = 1{,}000$ ms, and $t_{mut} = 20{,}000$ ms. On average, there have been 7 successful robots per successful run. These results show that CA is evolvable by using MARB controllers, and it has been achieved by an absolute number of 545 robots. However, there are no parameter combinations which guarantee a successful evolution, and even in successful runs on average more than two thirds of a population have not achieved CA behavior.

Tab. 4.3 shows how the successful runs are distributed among the tested values of each parameter separately. Some of the results indicate a tendency, in which direction the parameters should be shifted to improve the success rate. Particularly, the number of successful runs obviously increases with a larger mutation interval t_{mut}. A value of $t_{mut} = 5{,}000$ ms has yielded successful runs in 18 % of cases only, while a value of

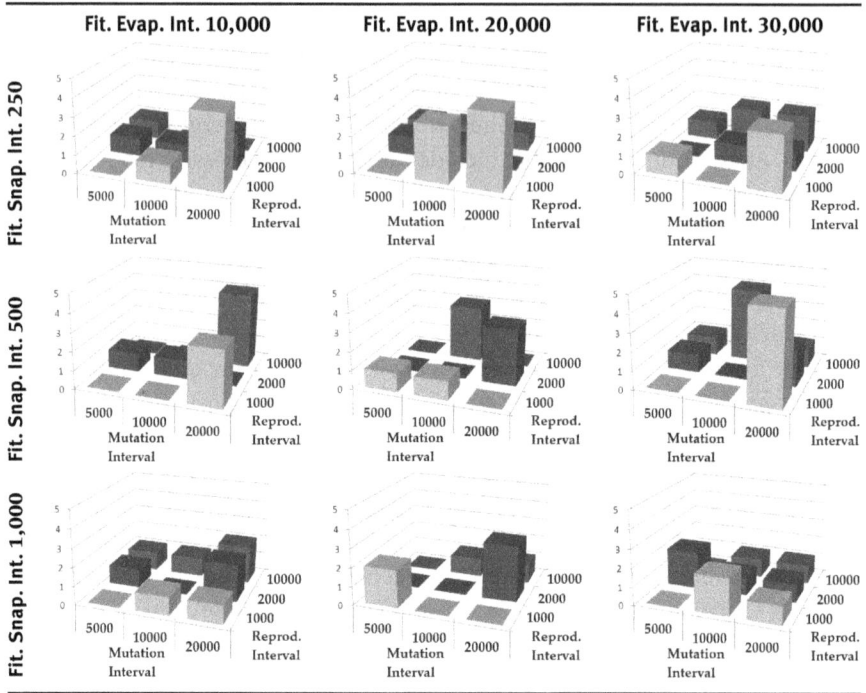

Fig. 4.10. Distribution of successful runs. For each of the 81 tested parameter combinations, the according number of successful runs is displayed. As 10 runs per parameter combination have been performed, the theoretically maximal column height is 10, however, 5 is the highest value achieved for any of the parameter combinations.

t_{mut} = 20,000 ms has yielded in 53 % of cases successful runs. Therefore, a further increase of the mutation interval seems to be a possibility to improve the results. A higher mutation interval, which is equivalent to a lower mutation rate, to some extent corresponds to a smoother mutation operator. The "smoothness" of the mutation operator is a topic further covered in Chap. 5. For the reproduction interval t_{rec}, no such statement can be derived from the available data. The values between 1,000 ms and 10,000 ms seem to be interchangeable without a significant effect on the results. For the fitness snapshot interval t_{snap}, it seems as if lower values gain better results, however, the differences are too small to allow for a significant statement. Similarly, higher values seem to be slightly better for the fitness evaporation interval t_{evap}. For more precise indications about individual parameter optimization, the data of this preliminary study is too sparse. Furthermore, due to interdependencies between the parameters, it may not be possible to find the perfect parameter combination by only optimizing each parameter separately. In Sec. 4.3, more experiments with larger sets of parameter combinations are performed using the suggestions derived here as a foundation.

Table 4.3. Distribution of successful runs for each parameter separately. The percentage values denote the amount of successful runs resulting from the respective parameter value. The most noticeable dependency is a rising success rate with an increased mutation interval t_{mut} indicating a potential to further improvement by increasing the mutation interval above 20,000 ms.

	Mutation Int. t_{mut}	Reproduction Int. t_{rec}	Snapshot Int. t_{snap}	Evaporation Int. t_{evap}
1.	5,000 ms: 18 %	1,000 ms: 37 %	250 ms: 37 %	10,000 ms: 31 %
2.	1,0000 ms: 29 %	2,000 ms: 29 %	500 ms: 36 %	20,000 ms: 30 %
3.	20,000 ms: 53 %	10,000 ms: 34 %	1,000 ms: 27 %	30,000 ms: 38 %

Evolved MARBs. Overall, 545, i. e., 2.6 % of all robots have been successful. Some have evolved the expected CA behavior, i. e., to move arbitrarily until an obstacle appears, then to turn until facing away from the obstacle, and then to continue moving around. Others have learned behaviors such as driving in circles or ellipses by oscillating between a *Move* and a *Turn* state, which matches the fitness function less accurately but still gains positive fitness. For some behaviors it is even by observation hard to decide what their basic properties are as they depend on specific environmental circumstances. For example, some robots avoid obstacles only when a specific sensor value constellation of the sensors from the back is received, which is very unintuitive to humans. For such behaviors a detailed analysis of the corresponding MARB has to be performed to detect the underlying processes. Moreover, a statement about the evolutionary circumstances which led to such behavior is even harder to achieve. Here, such an analysis has not been performed, and all objective statements about robot behaviors have been solely based on the respective fitness values. With that respect, some of the non-CA behaviors such as circle-driving may be interpreted as false positives according to their rating as being successful. However, it is more useful to interpret them as local optima on the way to CA, which should be overcome by increasing selection pressure.

Fig. 4.11 and Fig. 4.12 show two evolved MARBs, one of which performs CA behavior while the other one is driving circles. The MARB in Fig. 4.11 basically accomplishes a straight forward drive, except when sensing an obstacle, in which case the robot turns left as long as the obstacle is blocking the front, and continues driving when the front is clear again. Such a behavior seems reasonably close to a global optimum for the CA fitness function. The MARB in Fig. 4.12 performs a circle-driving behavior without sensing obstacles at all. However, due to the implementation of crash simulation, it turns out that a circle-driving robot finds, after a few initial collisions, enough space to drive nearly without any collisions and, therefore, is partially successful in the sense of the fitness function. Still, circle-driving is only a local optimum for the CA fitness function, as such a robot unnecessarily performs a turn every other simulation tick which leads to a loss of the fitness bonus for driving straight forward. This local optimum seems to have fairly stable properties in the evolutionary process of CA, and

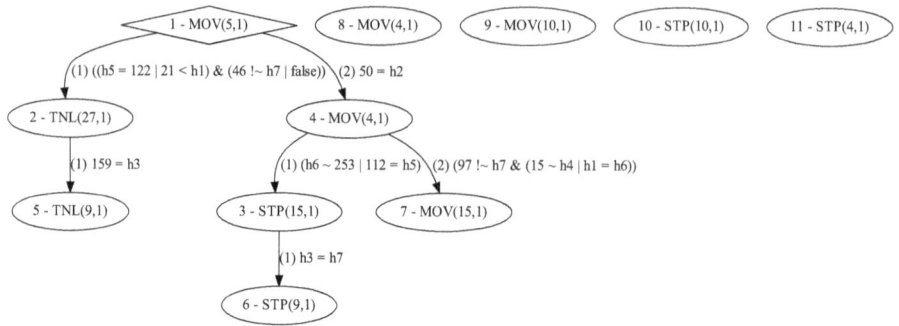

Fig. 4.11. Evolved MARB performing the expected CA behavior. The automaton has four unreachable states and three separate *Move* states as well as two separate *TurnLeft* states which are used to turn away from obstacles. This MARB has achieved a fitness of 730 in the final generation.

it occurs rather frequently even in the improved evolutionary setups described in the subsequent sections.

4.2.4 Concluding Remarks

Overall, the results show that the presented approach based on MARB controllers is in principle capable to evolve CA behavior. The parameters for this study have been selected rather arbitrarily and set to quite arbitrary values. Nevertheless, a parameter combination has been found, which has yielded successful runs in 5 out of 10 cases, and others have achieved 3 or 4 successful runs. Moreover, a high absolute number of robots has learned CA behavior within the runs. This indicates that further improvement can be expected when more specifically selected parameter combination are tested.

However, the experiments also uncovered some problems with the presented evolutionary setup. The process of reproduction and selection is based on the assumption that the robots are moving around in the environment. If too many robots are standing still, the rule of producing offspring always with the closest neighbor determines an "incestuous" mating with the same partner over and over again. As 72 % of the unsuccessful robots lack a reachable *Move* state, it can be expected that this problem has occurred frequently. A similar clustering problem arose in experiments with real Jasmine IIIp robots [100, 105], therefore, the movement of robots in the environment has to be considered a crucial factor of the presented approach. A higher mutation rate can lead to more diversity and consequently to a higher chance of *Move* states appearing in deadlocked populations, however, this idea conflicts with the result that a higher mutation interval should be considered. In the studies performed in the remainder of the chapter, this problem could be successfully reduced to a minimum by sub-

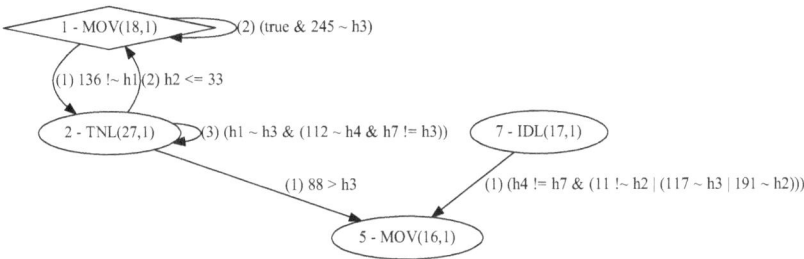

Fig. 4.12. Evolved MARB performing a circle-driving sub-behavior of the expected CA behavior. The MARB has one unreachable state (state 7) and basically switches constantly between a *Move* and a *TurnLeft* state letting the robot drive a circle. This MARB has achieved a fitness of only 132 in the final generation, losing the fitness bonus for moving forward any time when turning left without necessity and being punished for every crash with an undetected obstacle.

stantially increasing selection pressure using multi-parental reproduction. Another problem concerns the fitness function, which induces unexpected solutions such as driving circles without sensing obstacles. Although these solutions are partially successful in terms of the fitness function, they do not reflect the desired behavior. This problem, too, could be reduced by increasing selection pressure. In successful cases, circle-driving MARBs have been utilized as an intermediate step in the evolution of true CA behavior.

4.3 A Comprehensive Study Using the Examples of Collision Avoidance and Gate Passing

In this section, a set of experiments is described that build the main evaluation part of the evolutionary framework proposed in this chapter. Compared to the last section, the evolutionary setup has been improved in the suggested ways. However, these alterations include major structural changes such as using simulation ticks to trigger mutation, reproduction and other events, rather than real-world time. Therefore, the results can be compared to each other on a rough basis, when looking at successful vs. non-successful runs or specific evolved behaviors, but not on a detailed level, meaning specifically that quantitative fitness values cannot be compared between the two sets of experiments.

Table 4.4. Eight main sets of experiments. The fully factorial experimental design leads to eight main sets of experiments, for any of which the number of parents for reproduction has been varied from 1 to 10. Overall, 80 different experimental setups have been investigated. The table lists for any of the eight main sets the average percentages of successful runs and successful robots over all experiments with at least 2 parents; cf. discussion in Sec. 4.3.2.

	Memory Genome	Fitness function	Recombination	Succ. runs	Succ. robots
1.	no	CA	cloning reproduction	93.1 %	74.0 %
2.	yes	CA	cloning reproduction	99.5 %	87.6 %
3.	no	GP	cloning reproduction	87.2 %	73.6 %
4.	yes	GP	cloning reproduction	99.5 %	83.0 %
5.	no	CA	recombination *Cross*	92.7 %	75.3 %
6.	yes	CA	recombination *Cross*	99.6 %	88.7 %
7.	no	GP	recombination *Cross*	83.3 %	65.0 %
8.	yes	GP	recombination *Cross*	100.0 %	84.0 %

4.3.1 Method of Experimentation

The experiments in this section are closely related to the experiments described in the preceding section. In the following, the alterations to the experimental setup, compared to the preceding section, are described, assuming that all unmentioned aspects remain unchanged.

As described above, no real-world time measures have been used here, but all intervals are solely given in simulation steps. In addition to non-recombining reproduction, the recombination operator *Cross* as presented in Sec. 4.1.5 has been used in a subset of the experiments. The effects of recombining and non-recombining reproduction are compared with each other, and both are studied in combination with the use of the memory genome. Two target behaviors have been studied using the fitness functions for CA and GP. In detail, the following parameters have been varied in a fully factorial set of experiments:

- Utilization of the recombination operator *Cross* against non-recombining reproduction based on parental cloning;
- Utilization of the memory genome against not using the memory genome;
- Evolution of CA and evolution of GP;
- Variation of the number of parents ranging from 1 to 10, where 1 means no reproductive selection.

This leads to eight main sets of experiments, each of which is repeated with 10 different numbers of parents for reproduction; Tab. 4.4 summarizes the eight main sets of experiments. Thus, 80 different parameter combinations have been tested overall. Every combination has been simulated for 80,000 simulation steps, and every simulation has been repeated 26 times with different random seeds. Therefore, a total of $8 \cdot 10 \cdot 26 = 2{,}080$ simulations have been performed each of which lasts for approximately 45 minutes on average. This large amount of computation time has been handled using the EAS Framework and its connection to the JoSchKa job scheduling system [16], cf. Chap. 3. The remainder of this section describes the changes in the experimental setup and the evaluation of results in more detail.

The environment. All experiments have been performed on a rectangular field sized $1440\,\text{mm} \times 980\,\text{mm}$ (note the difference in size compared to Sec. 4.2). For CA the field has been empty in terms of walls or obstacles other than robots (Fig. 4.13a), for GP a wall with a gate in the middle has been placed in the field; the gate has an opening of $190\,\text{mm}$ (Fig. 4.13b).

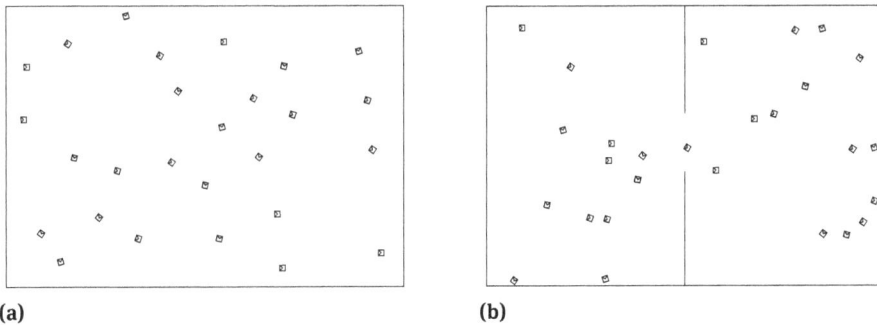

(a) (b)

Fig. 4.13. Depiction of experimental fields with robots drawn to scale. Experimental fields with 26 robots, set to random positions and facing in random directions as at the beginning of a run. (a) Field for CA. (b) Field for GP.

General settings. 26 robots have been placed randomly (in position and viewing direction) on the respective field. The initial MARBs have been empty, and the experiments have been run for 80,000 simulation cycles. This complies with a real-world driving distance of about 320 m for a robot which is driving straight forward or a real-world time of about 15 min. Mutation and reproduction have been performed every $t_{mut} = 100$ and $t_{rec} = 200$ simulation steps, respectively. For reproduction, the number of parents has ranged in $p \in \{1, \dots, 10\}$ for each of the eight main experimental setups. The runs with $p = 1$ serve as a control group, as they have not provided any selection pressure during evolution. The only way to direct evolution towards better fitness areas in these runs has been to mutate and reuse the memory genome, if activated. The fitness snapshot has been calculated every $t_{snap} = 50$ simulation steps and fitness evaporation has been performed every $t_{evap} = 300$ simulation steps dividing a robot's fitness value by $E = 2$. The interval for the memory genome has been set to 1,000 steps.

Definition of *success*. As before, robotic behavior is called "successful" (in relation to one of the two fitness functions used here) if it eventually leads to a positive fitness value when being executed in an environment. As argued before, for the described fitness functions, a negative or zero-fitness (after a proper time of execution) implies a non-adapted behavior with a high probability, while a positive fitness implies with a high probability an adapted behavior. Structural analysis of evolved automata with negative fitness at the end of the runs showed that in nearly all cases, they did not have both a reachable *Move* state and a reachable *Turn* state. This means that they were not capable of performing any non-trivial behavior. On the other hand, about 90 % of the robots with positive fitness had both state types reachable. Therefore, as before, an evolved robot is called successful if it has a positive fitness in the final population of a run, and an evolutionary run is called successful if at least one successful robot exists in the final population.

This definition of success is generalized to GP behavior here due to its simplicity, although it does not take into account the additional fitness bonus for passing the gate. As a consequence, robots performing CA in a GP experiment are also called successful. Therefore, for the GP experiments a further analysis on the MARB level has to be performed to decide if a successful robot is actually performing GP.

Negative fitness. As argued above, negative fitness correlates with trivial behavior, and it does not provide much information by the amount of negativity. Rather it distorts results if it is taken into consideration as in unfavorable situations robots can gain huge amounts of negative fitness in short time, e. g., due to a large number of collisions in a crowded corner. For this reason, as in the preceding section, in the following evaluations negative fitness is treated as zero.

Summary of changes. Overall, the study described in this section differs from that in the previous section in the following respects:
1. reproduction has been based on both simple cloning without recombination and recombination using the operator *Cross*;
2. the number of parents has been varied from 1 (control group) to 10;
3. experiments involving the memory genome have been added;
4. CA as well as GP have both been used as fitness functions;
5. both field types (for CA and GP) have been sized 1,440 mm × 980 mm (rather than 600 mm × 800 mm for CA in the previous section);
6. interval parameters such as the fitness snapshot interval t_{snap}, the evaporation interval t_{evap}, mutation and recombination intervals, etc. have differed in length and cannot be directly compared to the previous study due to the definition based on simulation time rather than real-world time;
7. overall, the experiments in this section have lasted for only 80,000 simulation steps, compared to approximately 2,000,000 steps in the previous study.

The runs with 2 parents, no memory genome, and no recombination operator, evolving CA, are supposed to reflect the previous experiments in Sec. 4.3.2). As the results show, despite the aforementioned slight differences in the experimental setup, the outcomes are quite similar and comparisons on a high level between the old and the new results appear to be valid.

4.3.2 Experimental results

With the new evolutionary setup, all sets of experiments greatly outperform the results presented before. There, 2.6 % of all robots and 11.0 % of all runs have been successful; here, 78.9 % of all robots and 94.4 % of all runs have been successful. Furthermore, the runtime has been drastically reduced from approximately 2,000,000 to 80,000 simulation steps (a reduction by a factor of 25). For the eight main sets of experiments, the percentages of successful robots and successful runs are given in Tab. 4.4; for these numbers, the control groups with only 1 parent for reproduction have been left aside. The reasons for this general improvement seem to be mainly

- the increased number of parents for reproduction, as two parents in the previous experiments seem to gain too little selection pressure, and
- the memory genome which prohibits the loss of already found good behavior.

The recombination operator seems to have different influences when used with or without the memory genome. In the following, a detailed analysis of the influences of these factors is given.

Influence of the memory genome. The memory genome seems to be largely responsible for the aforementioned general improvement. Fig. 4.14 shows the average percentage of successful runs, successful robots, and the mean population fitness in the last generation for the eight main sets of experiments separately. The runs with memory genome are plotted next to those without memory genome while the other parameter combinations are grouped along the X-axis. In every parameter combination, and for all three indicators, the runs with memory genome (dark bars/markers) outperform the runs without memory genome (light bars/markers). The plots in Figures 4.18, 4.19, 4.20 and 4.21 (see discussion below) show that this outperformance is present for all numbers of parents for reproduction, except for the control group with 1 parent.

When looking at fitness development during a run, the reasons for the good performance of the memory genome can be revealed. In the previous experiments, populations have tended to lose already found good behaviors by mutation. Therefore, studying only the last generations has not taken into account that good behaviors might have existed earlier in the runs. Also, such good behaviors might disappear before evolution has a chance to improve them, so they might not be able to reach their full potential. In the experiments with memory genome, fitness usually does not fall back permanently into negative values, once a good, robust behavior has been

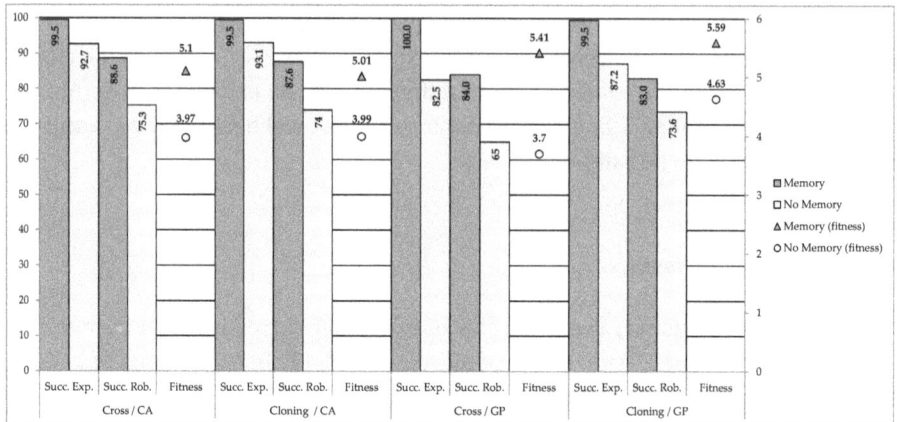

Fig. 4.14. Average results at the end of the eight main sets of experiments plotted for the memory genome. Average results in terms of percentage of successful runs, percentage of successful robots (both left axis), and mean fitness (right axis) at the end of the eight main sets of experiments. Runs with memory genome (dark) and runs without memory genome (light) are plotted next to each other.

achieved. Fig. 4.15 shows the average population fitness during a typical run with 6 parents where no memory genome has been used. At about 50,000 simulation steps, a high fitness is achieved, however, when approaching 60,000 simulation steps, fitness decreases drastically indicating that a formerly good behavior has been lost. Later a fairly high fitness is reached again, but it oscillates a lot and is not able to stabilize. In contrast, Fig. 4.16 shows the average fitness during a typical run with the same parameters except for an activated memory genome. A fairly high fitness is achieved at about 35,000 simulation steps and is conserved at a stable level until the end of the run.

A price that has to be paid when considering the usage of the memory genome, is a loss of diversity in the population due to the increased exploitation of already found behavior. While the memory genome is applicable to most decentralized evolutionary approaches, it might not be suitable for all of them as local optima might be preferred at the expense of a broad exploratory search. However, particularly in online-evolutionary approaches where a good solution has to be found as quickly as possible, and where the stability of behaviors is essential [105], the memory genome or a similar mechanism appears to be advisable.

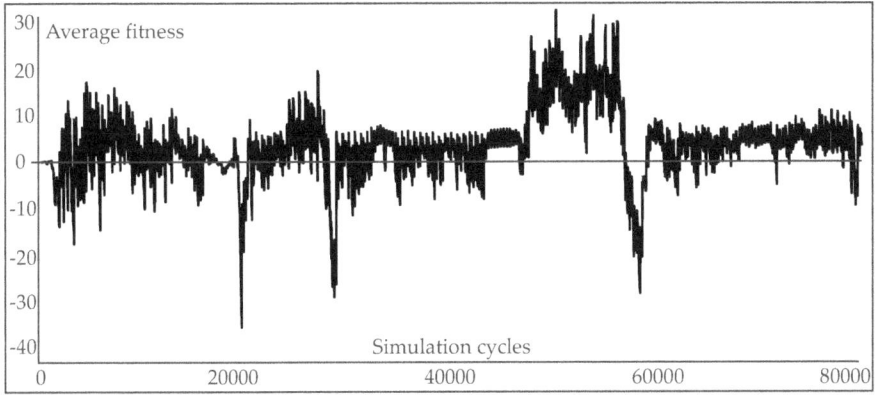

Fig. 4.15. Average fitness during a typical run without memory genome. The fitness has been averaged over all robots in every generation. The evolutionary setup includes 6 parents and no recombination during reproduction; the target behavior is GP.

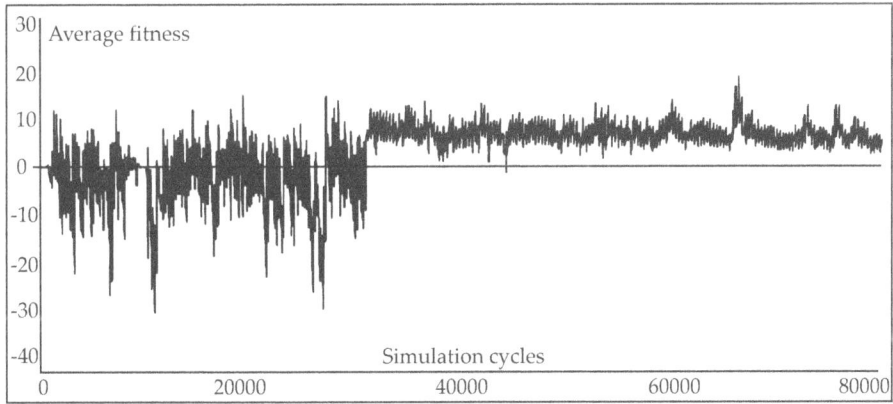

Fig. 4.16. Average fitness during a typical run with memory genome. The chart depicts a run with the same parameters as in Fig. 4.15 except for the usage of the memory genome.

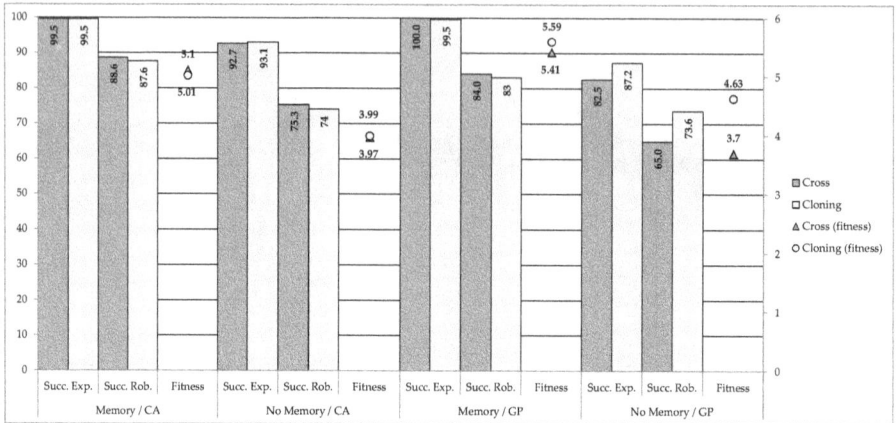

Fig. 4.17. Average results at the end of the eight main sets of experiments plotted for recombination. Average results in terms of percentage of successful runs, percentage of successful robots (both left axis), and mean fitness (right axis) at the end of the eight main sets of experiments. Runs with the recombination operator *Cross* (dark) and runs without recombination (light) are plotted next to each other.

Influence of the recombination operator. Fig. 4.17 shows the average percentage of successful runs and successful robots, and the mean population fitness in the last generation for the eight main sets of experiments. The chart depicts the same data as Figure 4.14, except that the bars are ordered to highlight the comparison between runs with cloning reproduction and runs with the recombination operator *Cross*. The chart shows that for runs without memory genome the cloning reproduction seems to outperform the recombination operator *Cross* – fairly clearly for the group evolving GP, and rather slightly for the group evolving CA. For runs with memory genome, the recombination operator *Cross* seems to slightly outperform cloning reproduction.

The results may be interpreted as follows. As the recombination operator *Cross* introduces additional diversity in the population, runs without memory genome are less capable of keeping good behavior and the probability increases that it gets lost. In runs with memory genome, however, the operator is capable to slightly improve the evolutionary effectiveness as already found good behavior is preserved and the additional diversity has a positive effect here. Therefore, in combination with the memory genome, the recombination operator *Cross* seems to provide a slight improvement of evolution. However, the enhancement is rather small, and purely statistical effects cannot be excluded based on the available data. The operator should be further studied and alternative recombination techniques might be considered before a conclusive statement is given.

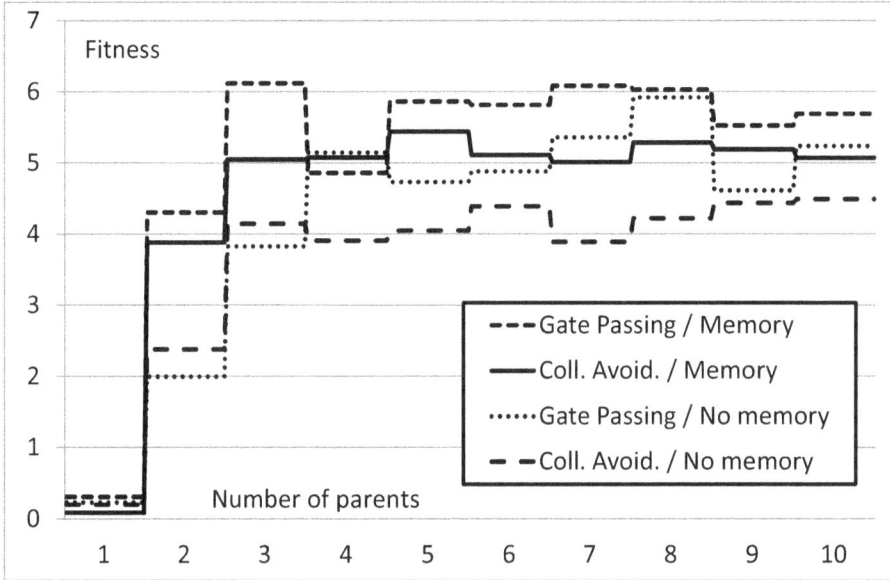

Fig. 4.18. Average fitness of robots in last populations of runs with cloning reproduction. The number of parents for reproduction is depicted on the X-axis. The Y-axis reflects the average fitness values in the last populations.

Influence of the number of parents for reproduction. Fig. 4.18 and Fig. 4.19 show the average fitness of robots and the average number of successful robots in the final populations for runs with cloning reproduction (i. e., main sets of experiments 1 – 4); different plot types denote the parameter usage (memory genome vs. no memory genome and CA vs. GP); the X-axis plots the number of parents for reproduction. Fig. 4.20 and Fig. 4.21 show the according data when using the recombination operator *Cross* instead of cloning reproduction (i. e., main sets of experiments 5 – 8).

As expected, runs with only 1 parent have not been able to achieve successful behavior in either case. Not even runs using the memory genome have reached a fitness significantly above zero despite the possibility to improve behavior by a simple random search using mutation only and storing the best genomes found by pure chance. However, as negative fitness values are treated as zero, the values for 1 parent are still slightly positive in all charts.

The performance of runs with 2 parents has been disproportionately better with the memory genome than without the memory genome (in relation to runs with 3 and more parents). This indicates that the memory genome is able to compensate for the lack of selection pressure, as observed in the preceding study, when allowing only 2 parents for reproduction. However, independently of the memory genome, both fit-

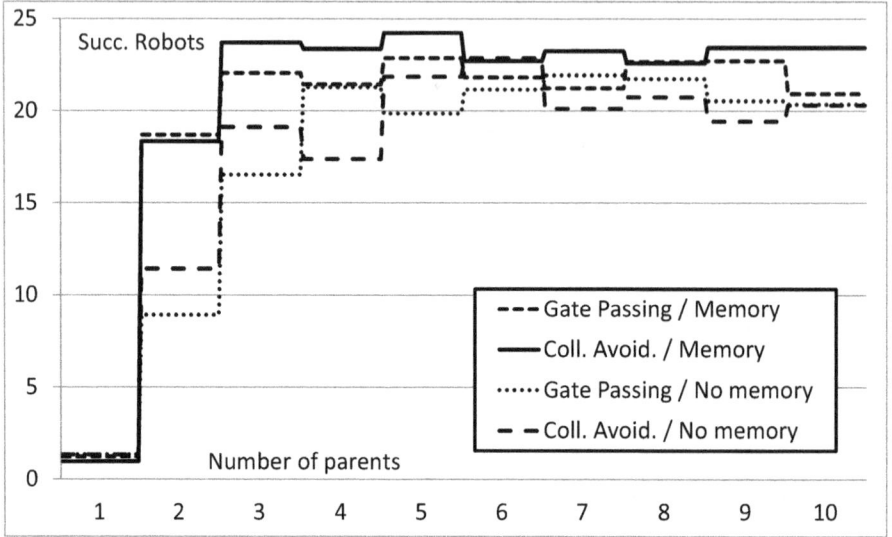

Fig. 4.19. Average number of successful robots in last populations of runs with cloning reproduction. The number of parents for reproduction is depicted on the X-axis. The Y-axis reflects the number of successful robots.

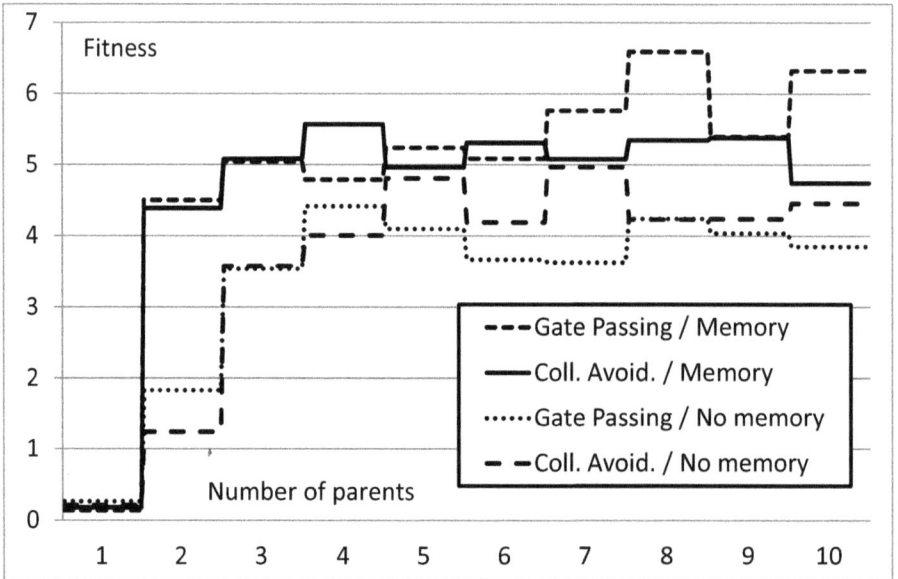

Fig. 4.20. Average fitness of robots in last populations of runs with recombination *Cross*. The number of parents for reproduction is depicted on the X-axis. The Y-axis reflects the average fitness values in the last populations.

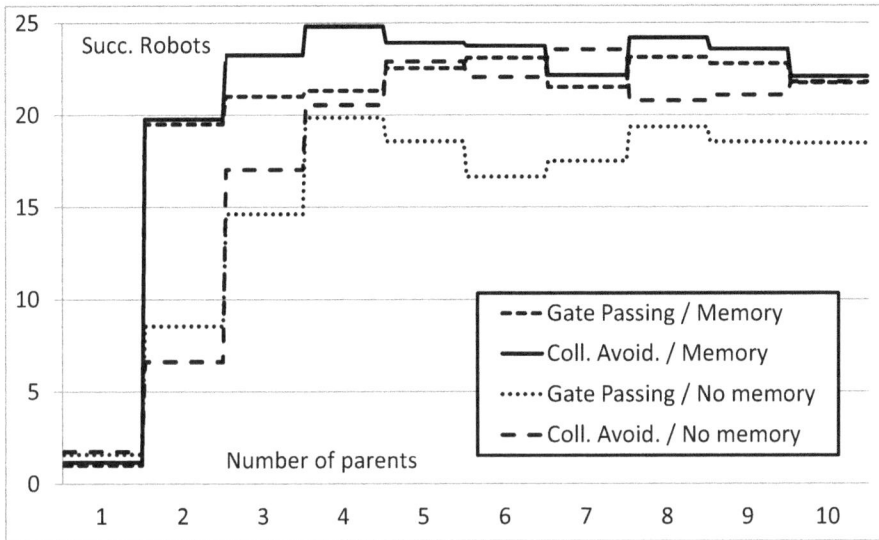

Fig. 4.21. Average number of successful robots in last populations of runs with recombination *Cross.* The number of parents for reproduction is depicted on the X-axis. The Y-axis reflects the number of successful robots.

ness and number of successful robots increase with the number of parents until an optimum is reached. The optimal number of parents seems to be between 4 and 7.

Over all numbers of parents in all parameter combinations, the runs with memory genome perform better than the runs without, which corresponds to the results presented above. Moreover, this result lines up in a series of results in EC that find elitist strategies to have positive effects on evolution if negative effects of the accompanying decrease in diversity do not prevail [201].

Overall, it can be concluded that selection pressure seems to be a key factor in improving the quality of evolution in the presented setup. As a decentralized multiparental reproduction technique is hard to realize on real robots in a reasonable way (see discussion in Sec. 4.4), particularly the memory genome seems to be a simple and effective way to provide the required selection pressure.

Analysis of evolved behaviors. Among the non-trivial and successful runs, roughly three groups of typical evolved behaviors have been identified:

1. *"Collision Avoidance"*: driving around arbitrarily (mostly straight forward) as long as no obstacle is in the way and bypassing obstacles otherwise;
2. *"Altruistic Gate Passing with Collision Avoidance"*: the same as "Collision Avoidance" as long as the gate is out of reach, identifying and passing it when it gets in sensor range, and continuing with "Collision Avoidance" afterward;

3. *"Egoistic Gate Passing"*: the same as "Altruistic Gate Passing with Collision Avoidance", except that in case the gate is found, a robot occupies it in a greedy manner and continues passing it back and forth blocking it for other robots; in some cases two robots share the gate, passing it constantly.

In each of the groups, behaviors of different complexity and robustness have been evolved. The assignment of evolved behaviors to the groups has been performed manually by observation and, for practical reasons, for a minor portion of the 2080 runs only. Therefore, no exact count of behaviors per group can be given. However, based on gate passage and collision counts, a weaker, somewhat obvious claim has been confirmed: in runs evolving CA only behaviors from the first group occurred while in runs evolving GP, behaviors from all three groups occurred.

Group 1 ("Collision Avoidance"): The behaviors from this group use obstacle avoiding strategies which range from very simple to highly adapted. Some simple MARBs perform just a circle-driving behavior without any obstacle detection, as described in Sec. 4.2. This can lead to a situation where nearly no collisions occur in the entire population, if every robot is following this strategy. However, the achieved fitness for this behavior is relatively low making it a local optimum in the fitness landscape. Other populations, on the other hand, have developed more sophisticated behaviors which implement different versions of what seems to be "true" collision avoidance. In a basic version, robots drive straight forward until an obstacle occurs, and then perform a "lazy turn", by oscillating between a *Move* and a *Turn* operation, always in the same direction. A further optimization of the turning behavior occurred in some populations where, e. g., robots learned to decide flexibly on which side to drive past an obstacle to decrease the required number of turning operations, thus increasing the number of fitness-gaining *Move* operations. Reducing the risk of collisions by still maximizing the number of *Move* operations is another example of further optimization that has been developed by some populations. There, robots learned to turn on the spot in very crowded regions, by remaining in a *Turn* state, while in regions with fewer obstacles a lazier and lazier turning behavior is performed by including more and more *Move* operations in the oscillation between *Move* and *Turn*. This can be interpreted as an adaptation to a dynamic environment as it is a reaction to the self-motion of obstacles, i. e., other robots. As an example, Fig. 4.22 shows a trajectory of a robot from group 1 in an empty field without other robots. The X marks the starting position, the robot is controlled by an evolved MARB. Fig. 4.23 shows a trajectory of the same robot in an environment with more obstacles. At the beginning it collides once with the small obstacle in the middle of the field, but afterward it is capable of driving around without further collisions.

In terms of the fitness function, the above described behaviors seem to be close to a global optimum of the CA fitness landscape. However, as CA is a subtask of GP, it is a local optimum in the fitness landscape for the experiments using a GP fitness

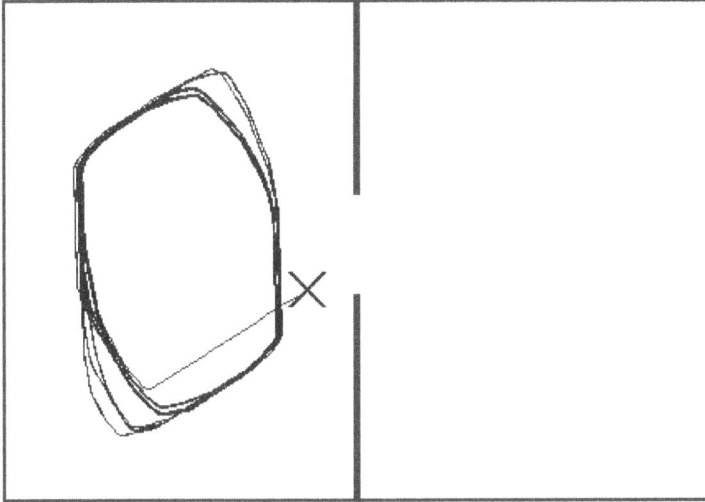

Fig. 4.22. Trajectory of an evolved robot from group 1 doing CA without passing the gate. The robot has been placed in an empty GP field at the position marked by an X and subsequently controlled by an evolved MARB.

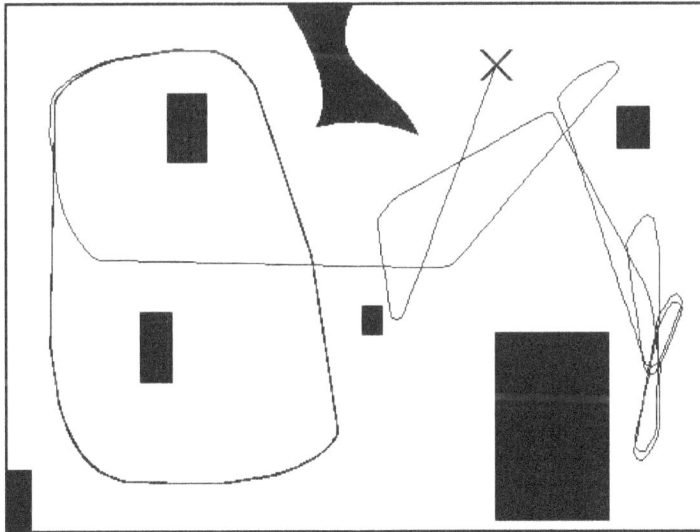

Fig. 4.23. Trajectory of an evolved robot from group 1 doing CA in a field with obstacles. Trajectory of a robot controlled by the same MARB as in Fig. 4.22 in a more complex environment. CA behavior typically generalizes well to different environment types.

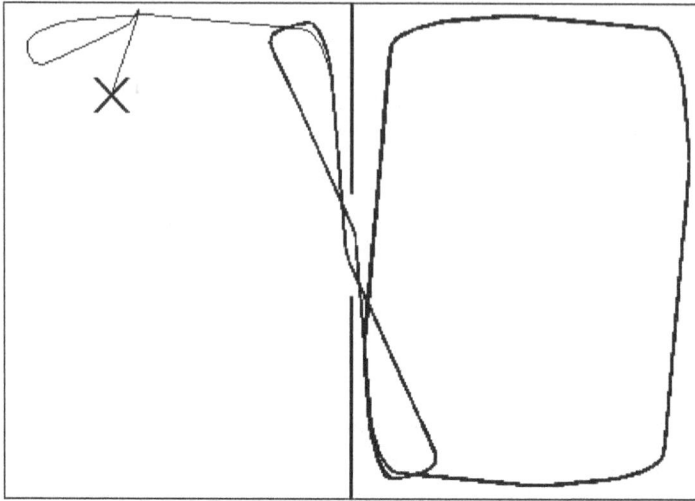

Fig. 4.24. Trajectory produced by a gate passing MARB from group 2. The robot either performs a looping behavior which involves passing the gate or circuits one half of the field without passing the gate. The trajectory has been produced by the MARB depicted in Fig. 4.27.

function. Therefore, many behaviors evolved in these experiments have been assigned to group 1.

An observation (originating from another, so far unpublished study) suggests that a drastic increase in robot density on the field (by leaving the other parameters unchanged) strongly promotes a higher sophistication of evolved CA behaviors. By evolving several hundred instead of 26 robots in the above-described field, the population tends to quickly converge to powerful CA behavior. This promotion seems to be greater than from just including static obstacles into the field and is overall unexpectedly strong. However, this observation is based on a small set of experiments and is, so far, a matter of conjecture.

Group 2 ("Altruistic Gate Passing with Collision Avoidance"): Behaviors in this group are the most effective ones in gaining fitness for the entire population. They all include an "altruistic" component which leads to a below-optimal individual fitness thereby facilitating a close-to-optimal global fitness. Fig. 4.24 shows the trajectory of a robot from this group. The robot is driving close to the wall avoiding collisions until it recognizes the gate and passes it. This does not happen every time it drives past the gate, but, due to a rather fuzzy recognition of the gate, only in about one out of two times which prevents the gate from being too crowded. After passing the gate, it performs a small loop on the other side and drives back through the gate. As the robot does not pass the gate every time, there is enough space for the whole population to profit from the behavior. Fig. 4.27 shows the corresponding MARB which is analyzed

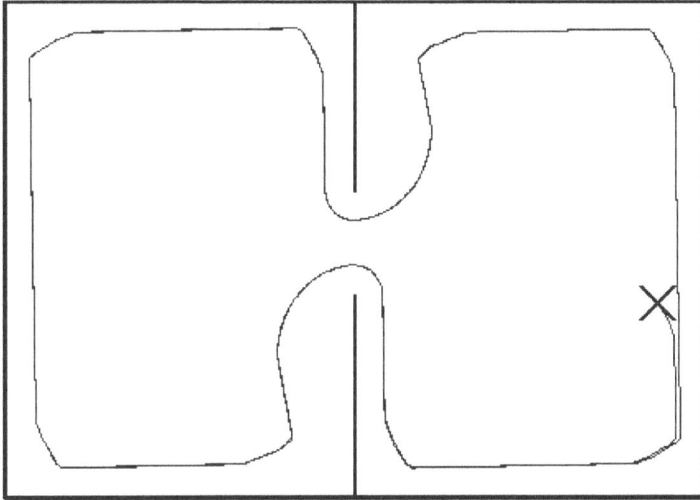

Fig. 4.25. Wall Following trajectory produced by a MARB from group 2. The robot performs a Wall Following behavior which leads to a circuiting through the entire field and two gate passages per round. Every robot from a population of 26 robots can benefit from this behavior as the gate is usually not crowded.

below. Another typical evolved behavior assigned to this group is Wall Following, i. e., a collision-avoiding behavior which additionally lets a robot stay close to a wall and follow its line. Wall Following robots circle in the environment and pass the gate twice during every complete round, a behavior every robot from the population can benefit from. Fig. 4.25 shows the trajectories of an evolved Wall Following behavior. A MARB evolved to perform Wall Following is analyzed in Chap. 5.

Group 3 ("Egoistic Gate Passing"): This group consists of behaviors which inhabit a rather undesired fitness niche exploitable for at most one or two robots at a time, gaining, however, great fitness values for them. Mostly, a mechanism has been evolved to recognize the gate and then to constantly pass it, back and forth. Fig. 4.26 shows a trajectory of a robot performing this behavior. In some populations, the gate passing robots interchange, in others, one or two single robots perform the gate passing while other robots are unable to recognize the blocked gate. In many of these populations, CA has not been evolved as a sub-behavior, as the constant passage of the gate gains high fitness by implicitly being safe from obstacles as long as no other robots initiate a crash. However, population fitness in this group is lower than in group 2. Depending on the application, such behavior may be adequate or, more likely, unintentional. However, as each robot individually struggles for high fitness during evolution, the behaviors from this group cannot just be considered local optima along the way to a "better" behavior from group 2. Rather, both groups can be considered close to a

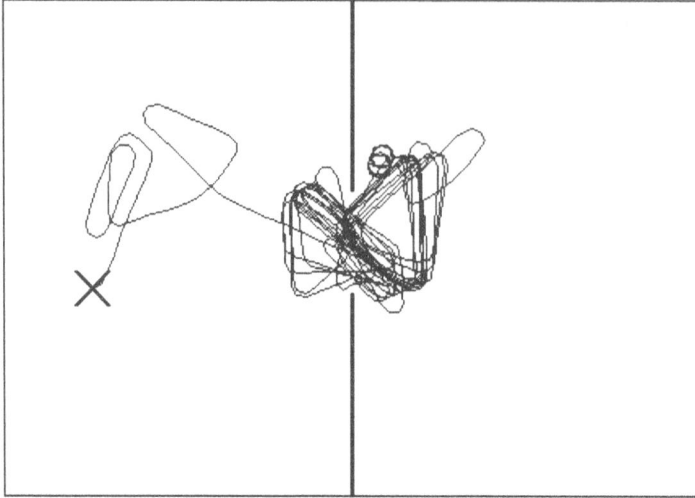

Fig. 4.26. Trajectory produced by an evolved MARB from group 3. While the robot constantly drives through the gate, it occupies the gate and blocks it for other robots which usually are not able to find the opening.

global optimum with equal justification. Which type of behavior finally occurs seems to depend on early stages of an evolutionary run; either the expected path is followed by first evolving CA and subsequently improving it, e. g., toward Wall Following; or an "egoistic" mutation in a specific situation leads to one robot gaining huge amounts of fitness and overruling other behaviors.

Analysis of the structure of an evolved MARB. Fig. 4.27 shows a MARB from group 2, evolved with the GP fitness function and performing the behavior depicted in Fig. 4.24. The MARB consists of six states, however, states 4, 5, and 6 are not reachable from the initial state. The corresponding transitions as well as the transition from state 3 to state 2 and the transition from state 2 to itself are inactive (the latter two due to the unsatisfiable conditions). From a semantic point of view, the automaton consists of 3 states and 5 transitions which are involved in the behavior. The transitions have fairly complex conditions which allow the robot to recognize and pass the gate. By the design of the mutation operator, all inactive states and transitions can be deleted within a small number of mutations, as they are not hardened. For example, the states without any incoming transitions can be deleted by a single application of mutation M_2. Using mutation M_7, the transition from state 3 to state 2 can be reduced to $false$ and deleted using M_5. Such inactive parts of a MARB build neutral plateaus for exploratory experiments that can either be involved in the behavior at some point or deleted eventually. In addition to the inactive parts, further neutral elements exist in the conditions; e. g., all constant parts ($true, false$) could be replaced by equivalent

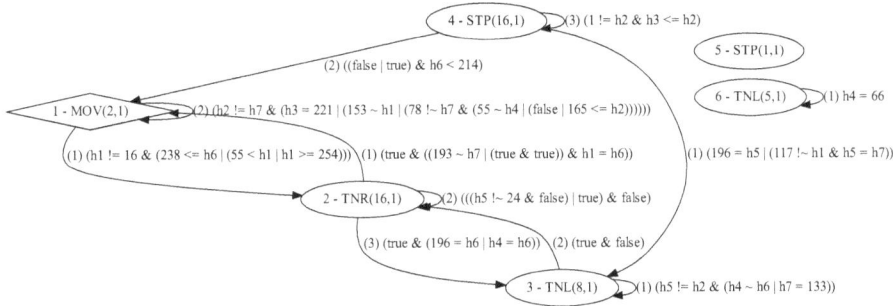

Fig. 4.27. MARB avoiding collisions and passing the gate. This MARB from group 2 corresponds to the trajectory in Fig. 4.24. The evolved MARB is depicted without any alterations, including unreachable states and inactive transitions.

simpler conditions. These elements, too, serve as neutral plateaus which may eventually be deleted by mutation if they are not used. In contrast, the hardened parts of the MARB, such as the interconnected states and the complex conditions, are unlikely to get removed or harmfully altered within few mutations.

4.3.3 Concluding remarks

It has been shown that an EE approach based on the MARB model is capable of robustly evolving CA in a variety of parameter combinations, and GP frequently in several variations. Overall, 78.3 % of the robots have achieved positive fitness (successful robots) and in 93.9 % of the runs robots with positive fitness have occurred in the final population (successful runs). This includes even the least successful parameter combination, excluding only runs with 1 parent for reproduction. The runs lasted for 80,000 simulation cycles which can be compared to 15 minutes in an experiment with real Jasmine IIIp robots.

The results show major improvements in terms of evolutionary success compared to preliminary results (cf. preceding section) which has been accounted to increasing the number of parents for reproduction and to using the memory genome as a decentralized elitist strategy. Both these methods increase selection pressure which seems to be a key factor of the presented approach. The recombination operator *Cross* seems to slightly improve evolutionary success when being used together with the memory genome and to decrease it otherwise. This can be explained by additional diversity which is introduced when using a recombination operator leading to an improvement only if the selection pressure is still high enough to exploit already existing good behavior in the population.

Due to a special design of the mutation operator, the complexity of the generated MARBs is intended to be adapted to the problem complexity during evolution; an analysis of evolved MARBs supports this proposition. While this does not necessarily mean that evolved MARBs have as few states as possible or few or simple transitions, it can be observed that states or transitions which are hardly involved in the behavior can be deleted, and parts of conditions which do not affect the behavior can be simplified within few mutations. Therefore, evolution can explore neutral plateaus which are only loosely connected to the evolved MARBs and flexibly alterable as long as they are not found to be useful.

An analysis of the resulting MARBs yields roughly three types of evolved behaviors: (1) "Collision Avoidance", (2) "Altruistic Gate Passing with Collision Avoidance", and (3) "Egoistic Gate Passing". There, from an applicative perspective, groups 1 and 3 represent a below-optimal behavior when evolved with the GP fitness function. However, while group 1 can be considered a sub-behavior of GP which can result in GP if further subjected to evolution, group 3 represents another type of solution to the problem posed by the GP fitness function. To avoid the occurrence of such behaviors, changes in the fitness function would have to be performed. Overall, behaviors from groups 2 or 3 have been evolved in about 45 % of the runs only, but virtually all runs can be considered successful in terms of the CA-based definition of success given above. Experiments with longer evolution time are expected to lead to improved results with respect to GP behavior.

4.4 Experiments With Real Robots

In this section, the experiments performed in simulation are transferred to the real-robot platform Wanda. The results have originally been published as a technical report [66], and further details regarding the implementation on real robots are given in a German diploma thesis [65]. As experiments with real robots are far more time-consuming than comparable runs in simulation, a significantly smaller number of experiments has been performed. However, insights from the simulation runs have been used at design time of the experimental setups to greatly reduce the number and ranges of parameters to be tested. By this means, a rather small number of runs has been required to find successful parameter combinations. The implementation on real robots has been kept as close as possible to the implementation in simulation. However, due to the constraints given by hardware and full decentralization, particularly the process of selection has been adapted, and the synchronized approach used in simulation has been exchanged by a completely concurrent approach.

4.4.1 Evolutionary Model

In the following, the evolutionary model as implemented on real robots is described. The main focus is on the adaptations performed on the evolvable controller model as well as on mutation and reproduction. Furthermore, some details about the implementation on Wanda robots are given. All aspects of the model that are not specifically mentioned can be assumed to be equal to the simulation approach.

Reproduction. In a decentralized scenario, reproduction requires robot-to-robot communication which is established by using the six IR sensors on the CTB of the Wanda platform. Before designing the reproduction process on a high level, the realization of communication on the level of IR signals has to be considered. An overview of the respective techniques is given in the implementation part below, more details are available in [65, 66]. Particularly, as IR communication is an error-prone technique, an error correction protocol has been proposed which is adapted to the small bandwidth available when using IR.

 A robot is allowed to communicate with exactly one other robot at a time to reduce communication complexity. Nevertheless, reproduction with $p > 2$ parents is possible by using a sequential method of collecting MARBs of other robots. There, a robot, called *collector* in this regard, communicates several times with different robots, storing all their MARBs with associated fitness values in its memory (the robots it collects from, are, from their own perspective, collectors as well and store MARBs from each communication for their own reproduction procedure). The actual reproduction is performed onboard of each individual robot after $p - 1$ MARBs have been collected. The collector builds a tournament with the collected MARBs using the according collected fitness values to determine the winner MARB. This MARB replaces the collector's previous controller.

 Due to the necessity of spatial proximity for communication via IR, it is a natural method to let the MARB collection take place anytime two robots happen to meet, i. e., enter each others' communcation radius. Therefore, reproduction is not synchronized, but emerges from the interaction of robots with their environment. Accordingly, the other parts of the evolutionary process are asynchronous as well, and particularly the evaporation part of the fitness function (cf. above) has to be carefully designed to guarantee comparable results at any time. The minimal time between two reproductions has been set to 100 cycles to prevent robots from repeatedly exchanging their MARBs in "reproduction clusters", cf. [105].

The memory genome. The memory genome is implemented by letting every individual keep in its storage the highest rated controller found so far including its fitness value. As before, in a constant interval it is checked whether the current fitness value is below a lower bound, and the current MARB is replaced by the stored MARB in that case. Specifically, every 6,000 cycles (cf. below) the following action is performed: the current controller is replaced by the stored controller if and only if the current fitness value is less than half as large as the stored fitness value.

Controller Model. The controller model used in this section is the MARB model as defined at the beginning of this chapter. While the overall model remains unchanged, a slight difference is given in the interpretation of movement commands. The first parameter par_1 of a MARB movement operation (cmd, par_1, par_2) with $cmd \in \{Move, TurnLeft, TurnRight\}$ does not indicate the distance to drive or angle to rotate anymore, but the driving or turning speed, respectively. This change, however, is structurally rather small as reduced speed can be established, for example, by oscillating between a *Move* and a *Stop* state, and, conversely, driving a certain distance or turning a certain angle can be established by visiting according states several times. Accordingly, the resulting MARBs mostly turned out very similar to the MARBs resulted from simulation runs.

Fitness. Some modifications to the above-defined fitness calculation are required to fit in an asynchronous scenario. Furthermore, the change in movement semantics, which include smooth wheel speed adaptation now, should also be considered. The modifications affect the fitness snapshots and the calculation of evaporation.

In the case of CA, a function of wheel speeds and the front distance to obstacles is used to calculate the fitness snapshot $snap_{CA}^{W}$ of a Wanda robot W. It is given by

$$snap_{CA}^{W}\left(t^{W}\right) = \frac{\left(v_{l}^{W}\left(t^{W}\right) + v_{r}^{W}\left(t^{W}\right)\right)d\left(t^{W}\right)^{2}}{200}$$

where $t^{W} \in \mathbb{N}$ denotes a discrete point in time, called the *(evolutionary) cycle*, measured by each individual robot W since the beginning of a run. It corresponds to the number of clock cycles given by the robot's operating system (flexibly adjustable in length, and set to about 100 ms per cycle, see below). Based on this notion of time, $v_{l}^{W}(t^{W})$, $v_{r}^{W}(t^{W}) \in \{-100, \dots, 100\}$ represent the left and right wheel speeds, respectively, at time t^{W}, and $d(t^{W}) \in \{0, \dots, 100\}$ represents the distance to the closest obstacle sensed by the front sensor at that time. As the wheel speeds always have the same absolute value (turning means setting both wheels to the same speed, but in different directions), only a forward motion has a positive contribution to the snapshot. Through the use of the square of the distance, smaller distances are rated superlinearly worse.

The snapshot of the GP fitness function consists of a CA part and an additional reward $g \in \mathbb{N}$ for driving through the gate:

$$snap_{GP}^{W}\left(t^{W}\right) = snap_{CA}^{W}\left(t^{W}\right) + \begin{cases} g, & \text{if gate passed since } t^{W} - 1 \\ 0 & \text{otherwise} \end{cases}$$

As before, fitness values are not reset after mutations, but expected to adapt to new behavior quickly. Nevertheless, it has to be assumed that fitness is delayed, particularly after mutations. Therefore, evaporation is required for deminishing influences

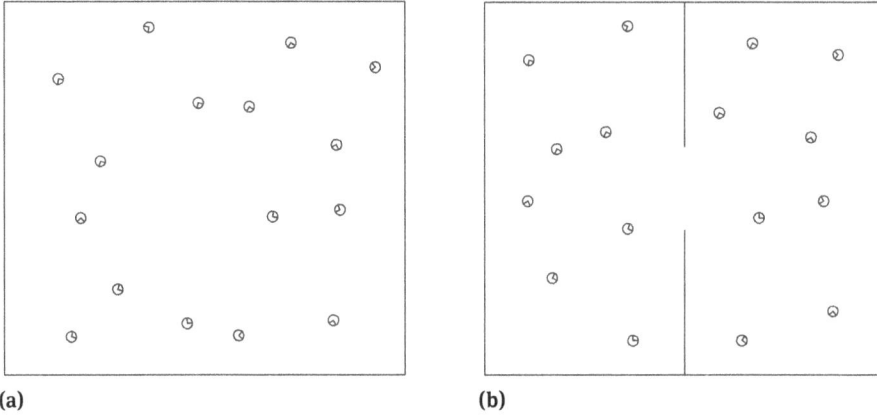

(a)   (b)

Fig. 4.28. Depiction of experimental fields with robots drawn to scale. Experimental fields with 16 robots set to random positions at the beginning of a run. (a) Field for CA; (b) field for GP.

of older snapshots over time. However, as fitness has to be comparable throughout the population at any time for reproduction purposes, the above-defined evaporation method has to be altered. It is replaced by an exponential moving average fitness calculation allowing for a smooth, continuous progression which yields values comparable with other robots at any time. The fitness values f_X^W with $X \in \{CA, GP\}$ are calculated by

$$f_X^W \left(t^W \right) = \alpha \cdot f_X^W \left(t^W - 1 \right) + (1 - \alpha) \cdot snap_X^W \left(t^W \right)$$

with $\alpha \in [0, 1]$ being the "smoothing" factor. By varying α, the influence of past behaviors can be regulated from high (close to 1) to low (close to 0).

Figs. 4.28a and 4.28b depict the fields used in the CA and GP runs, respectively. The robots are placed by humans at "random" positions facing in "random" directions (as far as humans are capable to do that) at the beginning of each run.

Mutation. The mutation operator is adopted without any structural changes from the operator M described at the beginning of this chapter. However, the following parameters have been altered. The mutation interval has been set to 50 cycles plus 1 % of the current fitness value, i. e.,

$$t_{mut}^W \left(t^W \right) = 50 + 0.01 \cdot f_X^W \left(t^W \right),$$

slowing down mutation for robots with high fitness values (thus providing a hardening mechanism at a "controller-as-a-whole" level – in addition to the above-described hardening of promising parts of a controller). The maximum step size used for mutations of labels (i. e., changes in state parameters or condition constants) has been set to 32 (rather than 5). Finally, the probabilities for choosing one of the atomic submutations have been changed [65].

Recovery from failures. Real-robot experiments include, compared to simulation runs, restrictions such as a longer run time, limited battery life and a limited life span of the robots [123]. Performing evolution in a swarm of robots can diminish these effects by introducing properties such as decentralization, robustness, flexibility and scalability [190]. For example, failures of single robots do not harm the entire experiment as other robots can adopt their function; and parallelism helps to reduce run time and to overcome short battery life [200].

Nevertheless, for the experiments described below, a recovery method has been found useful which stores the current state of a robot including its complete MARB and fitness history. The method has been implemented to allow for restoring a robot's state on demand, for example after a failure, and possibly on a different robot. Using this procedure, robots can be replaced after hardware failures or when the batteries are discharged.

While this is very convenient for making laboratory experiments more controllable, it affects the meaning of the results with regard to real-world evolutionary applications. More specifically, it has to be considered that the Wanda platform, in the exact state of development used here, is not actually capable to establish the runs described below in a completely unsupervised real-world scenario. However, the results remain valid when assuming a more robust robot platform with longer minimal battery life time. Therefore, the benefit of controlled laboratory experiments with a constant number of working robots during a predefined runtime has been preferred at the cost of the required human supervision. Nevertheless, the experiment lengths have not exceeded the theoretical battery life time of the robots. Of course, another benefit of the recovery function is that the MARB history can be used to analyze an evolutionary run in detail after termination.

Basic implementation. In the following, only a high-level description of the implementations on the Wanda platform is provided to allow for an understanding of the experiments described below. For an actual reimplementation or more detailed analysis, complete descriptions of the algorithms used on Wanda robots as well as several auxiliary tools can be found in [65, 66].

MARB Evolution has been implemented as a plugin for WandaOS in C++ requiring approximately 26 kB of program memory and 12 kB of RAM, by producing an occupancy rate of the processor of about 10 % to 15 %. To keep the implementation as flexible as possible, all variable parameters have been stored in a configuration file on a microSD card which is also used to store the recovery data. This allows for changes in the parameters, without changing the source code and re-programming the robot. In certain intervals, the configuration file is updated with the current state of the evolutionary process and saved to the microSD card for recovery. Therefore, a robot can be restarted and resumed after a failure at its last state before the failure, or replaced by another robot for that matter. In this sense the robot hardware itself is only a shell which can be exchanged to allow for a longer and more robust evolutionary process.

A log file containing a complete history of MARBs, their fitness values, information about mutation and reproduction as well as debugging information such as battery levels and communication failures is also written to the microSD card. There, MARBs are stored using the encoding described in Sec. 4.1.3.

Robot-to-robot communication is performed using a packet-based, connection-oriented and reliable protocol inspired by the *Transmission Control Protocol (TCP)*. However, due to TCP's significantly lower transmission speed and smaller packet size than required, only the basic ideas have been adopted. There are different packet types, some for the handshake phase, in which the IDs and the current fitness values are exchanged, and others for the MARB transfer phase. During communication, the robots stop their motion to avoid leaving the communication radius. Using a compression during transmission of MARB controller codes, one communication process takes 10–20 seconds, depending mostly on the MARBs' sizes. At each communication, the length of an evolutionary cycle is synchronized with the communication frames, meaning that the main loop is paused until a communication frame is completed. By this means, a cycle is set to a length of about 100 ms which roughly corresponds to a simulation step in the experiments described in the previous sections.

4.4.2 Method of Experimentation

A population of 16 Wanda robots is used for most experiments in the following. The size of the field has been determined according to the results obtained in the previous sections, and additionally according to the population size. The latter has been considered as IR communication tends to get worse with an increase in robot density. Good results have been achieved with five robots per square meter, resulting in a field size of $2,000 \times 1,600 \text{ mm}^2$ for 16 robots. This is, in relation to robot size, approximately the same field size as in the previous experiments, however, using a smaller number of robots.

For the GP experiments the field is divided in two equal parts by a wall with a gate sized 400 mm placed in the middle. One half of the field is illuminated using a video projector hanging from the ceiling. For fitness calculation, the robots detect passing the gate based on the brightness difference between the halves by using two light sensors on the OB. This sensoric aid is the only global information given to the robots in this section. It has been used for simplicity reasons as detecting the gate by the IR distance sensors only is, on the one hand, very complicated. On the other, and more importantly, detecting the gate via IR sensors is a major problem the robots are supposed to solve themselves. Providing the robots with a method for gate detection just to let them evolve one of their own seems rather tautological (although in theory, this would be the only way applicable to a real-world scenario where robots do not have access to global information – this reveals that GP is only a benchmark behavior which hardly justifies to be evolved in a real decentralized scenario; a more sensible approach, from

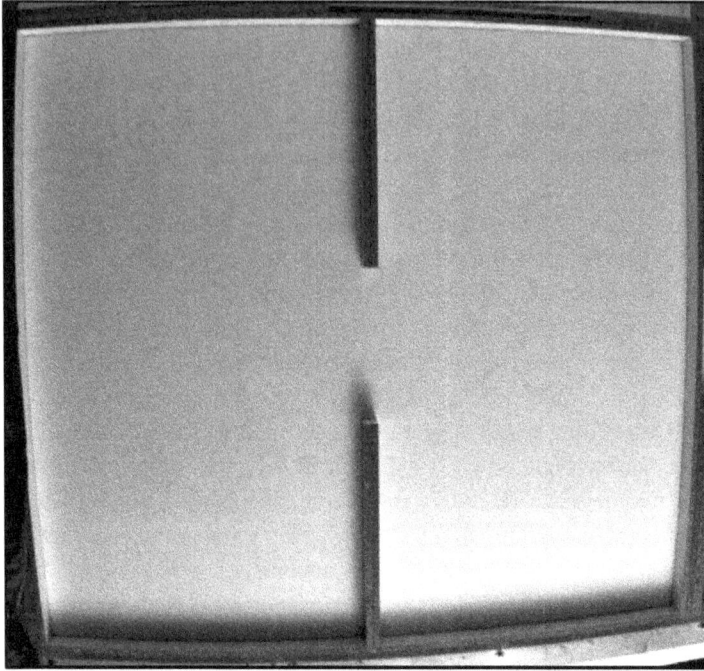

Fig. 4.29. Photography of the experimental field for GP. The photography has been taken by a high-resolution camera above the field. The camera produces a fish eye's effect that has to be removed before calculating trajectories.

an application perspective, might be to, inversely, use the evolved MARBs to create an IR sensor-based gate detection strategy; this, however, corresponds to a centralized, offline evolutionary approach).

As the robots' MARBs do not have access to the light sensors, the projected aid cannot jeopardize the results (although it has to be mentioned that a minor IR amount in the projected light may have been present, but no significant distortion of the IR sensors has been observed). All experiments have been recorded for later analysis with a high-resolution camera hanging above the field, and trajectories have been extracted automatically. For the generation of trajectories the software *SwisTrack* has been used [116]. Fig. 4.29 shows a photography of the experimental field for GP taken by the high-resolution camera.

Experimental setups. Overall, four sets of experiments have been performed, three of which used 16 robots and the above-described fields for CA and GP, respectively. To verify the advantage of parallelization of a decentralized and distributed approach, one set has been performed using only 8 robots in a $1,200 \times 1,300 \text{ mm}^2$ sized field. In addition to the variation in population size, the sequential reproduction operator has been varied with respect to the number of parents between 2 and 3. Overall, the following four experimental setups have been investigated:
1. evolution of CA, 16 robots, 2 parents;
2. evolution of CA, 16 robots, 3 parents;
3. evolution of CA, 8 robots, 2 parents;
4. evolution of GP, 16 robots, 2 parents.

For every setup, three evolutionary runs have been performed using different random seeds and slightly different initial conditions such as robot positions and angles. At the beginning of each evolutionary run, the robots have been distributed manually in the arena aiming at a preferably high distance from each other. To compare the experiments with the results achieved in simulation, a length of 80,000 evolutionary cycles has been used, which corresponds to a length of about four hours of experiment time. This has been calculated by the length of an evolutionary cycle in addition to the overhead produced by the runtime of the onboard implementations of the evolutionary model, communication etc. The overhead has been analyzed in preliminary experiments and accounts for about 40 % of the total time. (Due to this and to other aspects of the implementation – mainly the evolutionary cycle being set to a rather arbitrary length of 100 ms – as well as due to the fact that Wanda robots are slower than Jasmine-IIIp robots, this estimation is incomparably greater than the earlier estimation of 15 min real-world time corresponding to 80,000 simulation cycles.) After 80,000 evolutionary cycles every robot individually performs a controlled shutdown.

The smoothing factor of the fitness function has been set to 99 % for the experiments 1 – 3 and to 99.5 % for experiment 4. Thus, the influence of a snapshot falls within 250 cycles to 5 % of its original value in the CA experiments. In the GP experiments, this takes approximately 500 cycles, prolonging the influence of the reward for passing the gate. The reward has been set to a rather high value of $r = 400,000$ due to the exponential diminishing of fitness by smoothing.

As before, the success of a robot or an evolutionary run, respectively, is determined by the robots' fitnesses at the end of a run. A robot is considered successful if the average of its fitness values in the final minute of an evolutionary run has been positive. An evolutionary run is considered successful if it includes at least one successful robot. Furthermore, to gain more information about the behavioral qualities of a robot population, the development of fitness over time has been evaluated, and the time needed until the first robot of a population developed a "non-trivial" behavior has been studied. Non-trivial behavior has been defined to be given if the robot obviously reacts to environmental stimuli in a fitness-promoting way, for example, by

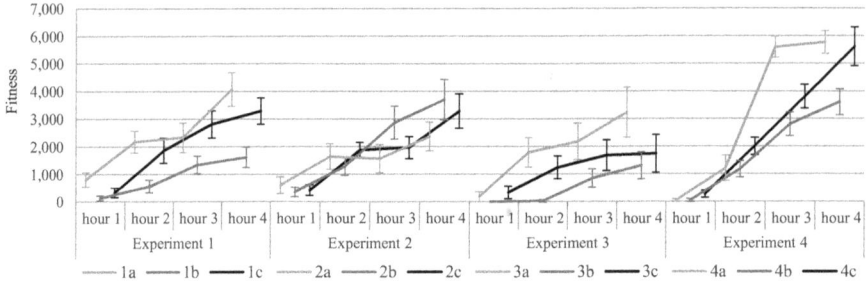

Fig. 4.30. Comparison of average population fitnesses over experiment time. The chart shows the development of the average population fitness for any of the 12 evolutionary runs on an hourly basis, categorized by the four main sets of experiments. The error bars denote the standard error.

turning away from a wall instead of crashing into it. The time to non-trivial behavior has been determined manually by observation of the video recordings. Additionally, the time from the first occurrence of non-trivial behavior to its spread through the entire population, determined by a positive fitness of all individuals, has been analyzed.

From a qualitative point of view, several of the best evolved MARBs of every evolutionary run have been analyzed manually and semi-automatically. For the GP experiment, the single best MARB of every run has been tested for ten minutes by copying it on eight robots and letting them interact with each other in an environment. The runs have been recorded and carefully observed, and trajectories have been extracted for all of them. Some notable examples are discussed below.

4.4.3 Results and Discussion

Overall, the following success rates have been achieved for the four main sets of experiments:

1. successful robots: 97.9 %, successful runs: 100 %
2. successful robots: 95.8 %, successful runs: 100 %
3. successful robots: 95.8 %, successful runs: 100 %
4. successful robots: 100 %, successful runs: 100 %

Furthermore, as in simulation, a strong correlation has been found between success and non-triviality showing that the given definition of success is reasonable. 100 % of the evolutionary runs and, on average, 97.4 % of all the robots have been successful (with a standard error of 1.4 %). The development of the average population fitnesses in the individual runs show, despite their small number and the influence of randomness, consistent results, cf. Fig. 4.30. In almost every run (except 2a), the average population fitness constantly increased on an hourly basis. The runs of set 4 have overall

Table 4.5. Periods to non-trivial behavior occurrence and establishment in the population. The first row shows the time of first occurrence of non-trivial behavior. The second row shows the time of establishment of non-trivial behavior in the entire population. The third row shows the time from the first occurrence of non-trivial behavior to its establishment in the entire population.

	Experiment 1	Experiment 2	Experiment 3	Experiment 4
First occurrence	00:42:20	00:14:00	01:02:20	00:52:00
Established in population	01:13:20	01:05:00	01:22:40	01:18:20
Time in between	00:31:00	00:51:00	00:20:20	00:26:20

the highest fitness values due to the additional fitness bonus given for gate passages. The runs of set 3 have the lowest fitness values which can be attributed to the smaller population size.

Influence of the number of parents. The study of the reproduction operator with respect to different numbers of parents, examined mainly in experiments 1 and 2, leads to no clear result. The fitness values in the runs with three parents are not significantly different from that in experiments with two parents. The rate of successful robots is even slightly higher in the experiment with two parents. A major reason for this result, which is opposed to the results from simulation, seems to be the more than twice as large reproduction interval in the runs with three parents compared to the runs with two parents. This blowup in reproduction time is caused by the sequential collection of two parental MARBs, and it would be even higher if more than three parents were chosen. This is also reflected by the time to establish non-trivial behavior in the entire population which is 65 % longer with three parents than with two parents, cf. Tab. 4.5. However, the higher selection pressure in the experiments with three parents appears to partially compensate for this disadvantage, since the fitnesses of experiments 1 and 2 are similar. To determine the answer to the question of a suitable number of parents, additional repetitions of the experiments are required. Overall, the results suggest that an increase of selection pressure achieved in simulation by increasing the number of parents cannot be established in a straight forward way on real robots.

Influence of parallelization. Experiments 1 and 2 have been significantly better in terms of fitness achieved and time required for achieving non-trivial behavior than experiment 3. The period until the appearance of first non-trivial behavior could be overcome significantly faster with sixteen robots than with only eight robots. Therefore, parallelization obviously gained the expected improvement for the evolutionary process. This is not at all surprising and a further discussion (as provided in [65, 66]) is omitted here.

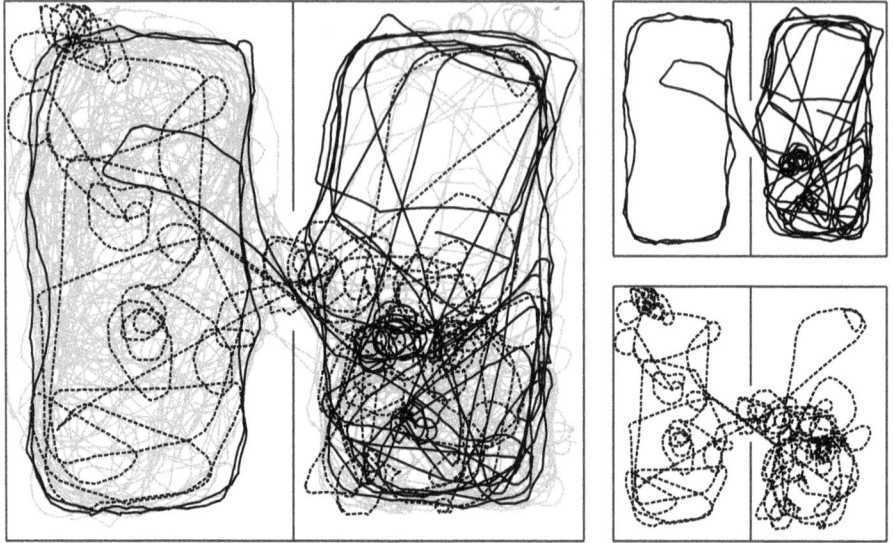

Fig. 4.31. Trajectories of an evolved GP behavior traced for 10 minutes. Left: trajectories of a population of eight robots (six suggested by gray solid lines, two emphasized in black – solid vs. dashed), all controlled by the same evolved MARB for GP behavior depicted in Fig. 4.32; right: the two trajectories which are painted in black on the left, depicted in separate fields.

Influence of the memory genome. Overall, in 64 % of all cases (with a standard error of 2.9 %) the memory genomes has been used to override the controller of a robot during the checking phase every 6,000 cycles. This confirms the neccessity of an elitist strategy as every time the memory genome is used, a major drop in fitness is prevented. The high utilization rate can also be an indication that the checking interval is too long. As no control group experiments without the memory genome have been performed, a final statement about its impact on real robots cannot be given. However, there is a strong indication that the memory genome has had a major impact on evolutionary success. Nevertheless, it has to be kept in mind that the memory genome, as every elitist strategy, reduces population diversity which can have negative effects on the evolutionary search.

Evolved behaviors. All experiments and many of the evolved MARBs have been observed and analyzed in detail. In all CA experiments, an obstacle avoiding behavior has been found at some point during a run, and particularly at the end of all runs some version of CA has been performed by a major portion of the robots in all evolved populations. In the more complex GP case, the robots of all three populations have found a simple way of following a wall (cf. Fig. 4.31). This behavior has already been discussed as a possibility to solve the GP task in a nearly optimal way, cf. Sec. 4.3.2 and other publications [103, 104, 126]. In the results here, the evolved Wall Following

(1) (50 != h2 & (22 >= h2 | 84 = h2))

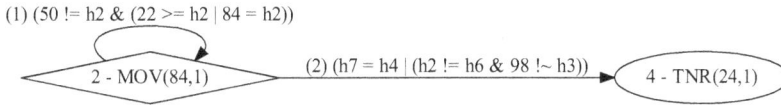

2 - MOV(84,1)   (2) (h7 = h4 | (h2 != h6 & 98 !~ h3))   4 - TNR(24,1)

Fig. 4.32. Simple evolved MARB for GP behavior. MARB controller for GP behavior corresponding to the trajectories in Fig. 4.31. Unreachable states have been omitted for better visualization. The initial state 2 has two outgoing transitions. Transition 1 is *true* as long as sensor h_2 signals free space in front of the robot. Transition 2 is almost always *true*, therefore, it is chosen most of the times when transition 1 is not *true*. This leads to a CA behavior of driving straight forward as long as no obstacle is ahead, and turning right if an obstacle appears. However, as the trajectories show, the MARB is additionally capable of sensing and actively passing the gate, a behavior that is encoded in the rather unintuitive subparts of the conditions.

is either relatively slow or the gate is occasionally missed (which possibly can be attributed to the influence of the CA part of the fitness function). In both cases, the gate is usually not crowded allowing for the whole population to benefit from the behavior.

Fig. 4.32 shows a MARB controller, evolved during a GP experiment, which is very similar to typically evolved CA controllers. Therefore, minor changes to a MARB performing CA can lead to a MARB which senses the gate and temporarily interrupts the CA behavior to perform a gate passage (note, in contrast, that human-designed solutions to the gate detection problem tend to be much more complicated and often less robust). This fact supports the assumption that CA can be used as an intermediate step in the evolution of GP.

The average time between two gate passages for this controller has been about 280 seconds which is very high when considering that the influence of the reward in a robot's fitness is lost after 500 evolutionary cycles, i. e., about 70 seconds when including time lost by technical delays (see above). Therefore, by this method of reward a gate passage cannot account for a permanent increase in fitness, and the main impact of fitness on selection appears to be the CA part of the fitness function. A higher reward should be considered in future experiments to yield a longer-lasting effect on selection. Nevertheless, a Wall Following behavior as a close-to-optimal GP behavior has been successfully evolved in all these runs, although a more stable gate detection would have been desirable.

Further observations. The hardening of those parts of a MARB which are highly involved in the behavior, by means of mutation and selection, has been found to work as predicted. A minimally required structure for a CA behavior, namely a reachable *Move* and a *Turn* state, are present in virtually every investigated individual. When removing unreachable and inactive (i. e., loosely connected) parts of evolved MARBs, most of the resulting MARBs are slim, particularly in the CA runs often having two states only. This indicates that an adequate search space complexity has been found by evolution.

The recovery feature has been used frequently and found greatly important for stable experiments. Otherwise, a much shorter experiment time would have been required, or many robots would not have reached the end of the evolutionary run.

Beside explicit selection based on the fitness functions for CA and GP, respectively, indications for implicit environmental selection have also been found occasionally. For example, robots located in a cluster of other robots often reverse after driving a short distance and return to the cluster to perform reproduction. This is from a controller point of view (in terms of "selfishness of a controller" from a game-theoretical perspective [34]) a successful way of establishing a stable behavior which can hardly be destroyed if it is present in all the robots of a cluster. Another example is the implicit selection toward shorter controllers. Shorter controllers have a higher chance of being passed on to other robots due to the lower probability of communication faliures. Particularly runs from experimental set 4 provide evidence for this type of selection. The effect of a weak pressure toward small controllers may well be desired (and is part of the intention behind hardening), while a clustering including repeated reproduction with the same "undeveloped" robots is rather undesired. In any case implicit selection cannot be neglected and has to be considered when setting up experiments. The impact of implicit selection on evolution is studied in more detail in Chap. 6.

Finally the three criteria for evaluating evolutionary experiments with real robots, proposed by Walker et al. [199], can be applied to the results:

1. "Influence of the evolutionary process on the task itself": As 40 % of the time robots are busy with technical aspects of the evolutionary process such as IR communication, a further improvement on the technical side is desired. The encoding of MARBs, as presented in Sec. 4.1.3, has not primarily been designed to be space-efficient, but rather to easily support mutation; there a compression algorithm might help saving memory and communication time. The neccessity to stop for communication occasionally leads to clusters of robots, which dissolve slowly and promote the cluster-based selection mentioned above. A more sophisticated reproduction procedure might help preventing this problem.

2. "Adaptation speed": An evaluation of adaption speed has only been performed by identifying the first occurrence of a non-trivial behavior; further nuances seem difficult to recognize. In the experiments with sixteen robots, an average of 30 minutes has been found to be required for this purpose, about an hour in the experiments with eight robots.

3. "Continuous improvement of behavior": As described above (and as Fig. 4.30 shows), the fitness continuously increased in all evolutionary runs which correlates with an improvement of behavior.

4.4.4 Concluding Remarks

A successful completion of the experiments for CA and GP confirm that it is principly possible to transfer MARB evolution as performed in simulation at the beginning of this chapter to the Wanda robot platform. The challenges presented above can be overcome by the proposed counteractions. A long evolution time can be shortened by distribution on a higher number of robots, and limited battery life as well as robot failures can be compensated by using a recovery method. The latter is, however, feasible in laboratory experiments only and has to be disabled in real-world scenarios. In all experiments, a positively rated behavior has, once discovered, been distributed throughout the population and in most cases not been lost again. Average population fitness increased continuously on an hourly basis in all cases, leading to a fairly effective version of the desired behaviors in the final populations. A positive effect of parallelization has been detected as an increase in population size has led to a super-proportional increase in average population fitness and speed of evolution. No final statement about the optimal number of parents for reproduction can be made. With the presented sequential version of the reproduction operator, a positive effect of increasing the number of parents from 2 to 3 has not been observed. Rather, the advantages and disadvantages between 2 and 3 parents appear to be in a balance and yield similar results.

In future work, further experiments should be performed to reduce the statistical uncertainty in the results and to explore new parameter setups. Following the successful results in the generation of CA and GP, they should be focussed on more complex tasks, that, for example, more inherently require the cooperation of multiple robots. Furthermore, the technical parts of the implementation can be further optimized. This concerns, above all, IR communication which currently particularly slows down the evolutionary process by requiring about 25 % of evolution time. By replacing IR communication with communication based on ZigBee, which is already available on the robots, the communication time could be majorly reduced. The available bandwidth of 250 kB/s is theoretically sufficient to allow for more than 1,000 robots exchanging their controllers every 10 seconds. However, due to the complicated implementation that requires an IR part for finding nearby robots and a ZigBee part for the actual controller exchange, this has been omitted so far. With the increase in communication speed, a complete interchange of controllers in a tournament could be easily established which would also facilitate the implementation of a recombination operator.

4.5 Chapter Résumé

In summary, it can be concluded that the proposed decentralized online-evolutionary approach using MARB controllers works well and is capable of robustly evolving simple behaviors such as CA, and frequently behaviors of intermediate complexity such as GP in simulation and on real robots. While these behaviors are comparable in their complexities with behaviors evolved in current literature (cf. discussion in Chap. 2), truly complex behaviors have not been evolved so far making this the most important goal for future research. A slightly more complex behavior for that matter is Wall Following as it is more specific in its behavioral demands. It has been evolved occasionally using the GP fitness function. Other well-known benchmark behaviors are Object Recognition, collaborative behaviors such as Object Transport, and similar behaviors as listed in Chap. 2. However, these behaviors are still far from what is imagined to be truly complex. Complex behaviors should involve sophisticated decision-making in dynamic environments and under varying circumstances, possibly by executing a task that has several objectives to consider. It is until today, from the results obtained here and in ER research altogether, still questionable if an approach using only classic evolutionary operators by itself can evolve complex behavior from scratch. It might well be that more sophisticated approaches are required that, for example, combine several different ideas from machine learning in a smart way, or that project the model that evolutionary biologists are still deciphering from nature more accurately on artificial evolution. The latter idea is followed in the next chapter; there an evolutionary model is proposed which mirrors the alongside evolution of the GPM with the evolution of phenotypic traits, as observed in nature, on ER.

5 Evolution and the Genotype-Phenotype Mapping

Preamble: *Recursion is a powerful principle – frequently applied in mathematics and the natural sciences as well as in very remote fields such as arts and literature. It describes objects or processes that contain one or more similar, often in some sense smaller version(s) of themselves inside of them. In his famous book "Gödel, Escher, Bach: an Eternal Golden Braid", D. R. Hofstadter calls recursive processes "strange loops" if they occur under non-intuitive circumstances [76]. Strange loops, while being somehow mystical and hard to understand, often harbor complex and in various ways valuable properties. The recursive process proposed in this chapter for the purpose of evolving genotype-phenotype mappings might appear strange at first. The idea, however, is quite natural at second sight and it proves powerful in making an evolutionary run adaptable to individual properties of a given search space.*

Of all controllable aspects of an evolutionary run, the *evolutionary setup*, i. e., the combination of an initial configuration of a run and the according evolutionary operators, is for many ER scenarios of highest influence on its success. In a strict sense, missing an adequate evolutionary setup is at the core of the bootstrap problem – as long as the overall evolutionary framework allows for an adequate setup for a given target behavior [56, 141]. As different target behaviors usually require different evolutionary setups, the procedure of finding a good setup has to be performed whenever a new behavior is being evolved. Furthermore, slightly differing evolutionary setups may lead to highly different outcomes (cf. experimental evaluation in Chap. 4, particularly Secs. 4.2 and 4.3) which makes many classic methods of experimental design rather useless [131]. If the environment changes dynamically, it is even more challenging to provide a setup which remains adequate during the whole process. The high effort of initiating a successful evolutionary run is a drawback for virtually all current ER approaches.

On the other hand, natural evolution has managed to evolve a huge variety of organisms of the highest imaginable complexities which continue to adapt to their ever-changing environment. The original evolutionary setup in nature is given by the laws of physics and the available chemical elements on earth, and it does not provide specific target constraints for organisms to evolve toward (if any, such constraints exist below the organism level in terms of producing preferably many copies of inheritable material in cells [34, 125]). Therefore, natural evolution is in substantial regards different from most artificial evolutionary approaches and to a great extent incomparable to the abstracted evolutionary setups used in ER. However, when looking at natural evolution from a higher level, any particular population of organisms does have specific desirable target properties such as the ability to find food or to escape predators. Furthermore, the individuals of a population belonging to the same species have common genetic configurations controlling, to some degree, mutation and recombination as well as how selection is performed. At a high level, this can correspond to an evolutionary setup in ER.

In nature, however, these high-level evolutionary setups are not fixed. Rather, past mutation, recombination and other factors have shaped the current evolutionary setup of a population, i. e., present mutation, recombination etc. along with the phenotypic traits of the individuals. There, mutation and recombination remain fairly stable at the genotypic level as they are mainly predetermined by external forces (such as cosmic radiation for mutation). Conversely, their effects at the phenotypic level are flexible and can radically change in the course of evolution. For example, mutation today works completely different in humans than in fruit flies in terms of how a specific mutation at the genotypic level affects changes to the phenotype [52].

The reason for this is that the original evolutionary setup in nature, i. e., the physical environment, is flexible enough to allow for differing high-level evolutionary setups to evolve in different populations. One major characteristic of this flexibility is nature's capability to change the way how genotypes are translated into phenotypes, i. e., the GPM. In nature, the genetic code, stored by the DNA in higher organisms, includes decoding instructions for the ribosomes which perform the translation process from DNA to proteins in a cell, i. e., the first step in the development of phenotypic traits. Therefore, the DNA contains not only phenotypic information (if such a specification even makes sense on this level), but also information about how it is to be translated into a phenotype. This makes the GPM subject to mutation and recombination, and, in the long term, to selection. The latter is a consequence of the effects of mutation and recombination becoming alterable on a phenotypic level by changing the GPM, cf. Fig. 5.1. Altering the GPM leads to changes of the effects of mutation and recombination which, in turn, can lead to improved phenotypic traits by supporting beneficial and avoiding harmful effects of mutation. It is important to bear in mind that the mutation and recombination operations on the genotype level are not affected by this process. Rather, the same mutation and recombination operations on a genotypic level can lead to different changes of a phenotype depending on the underlying GPM. A GPM which is subject to mutation, recombination and selection is called *evolvable GPM* in the following. (In fact, natural evolution establishes what is called a "completely evolvable GPM" below; however, the natural example should not be overstressed at this point and, rather, the focus returned to the presented ER approach.)

The next section gives a high-level overview of the presented approach, keeping the perspective close to the natural example presented above. In Sec. 5.2, the FSM-based approach for accomplishing an evolvable GPM is defined including the according evolutionary operators and search spaces. In Sec. 5.3, the results of a comprehensive three-part study are described and discussed showing the applicability and several evolutionary effects of the proposed approach. Sec. 5.4 concludes the chapter and particularly presents several ideas for future studies concerning the solution of some remaining problems.

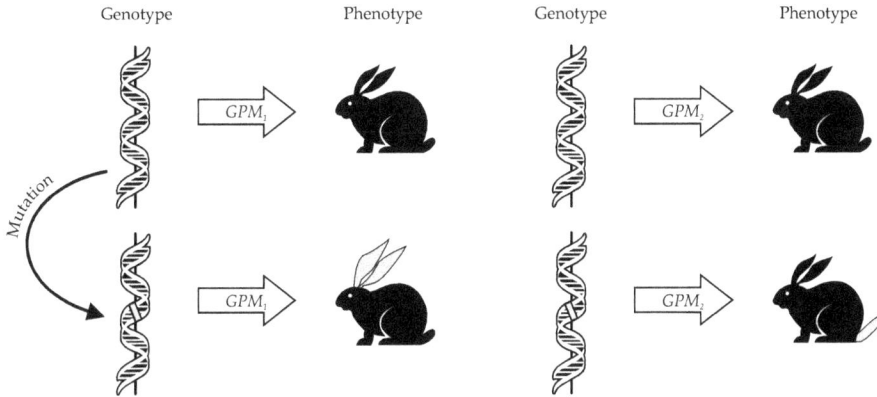

Fig. 5.1. Influence of the GPM on the effects of mutation. The picture shows a (strongly oversubscribed) sketch of two organisms, top left and right, at an equivalent stage of phenotypic development, both exposed to the same mutation. The left organism's mapping from genotype to phenotype is given by GPM_1 which leads to enlarged ears as a consequence of the mutation (bottom left). By the same mutation, the right organism's mapping GPM_2 leads to an enlarged tail (bottom right). Note that in nature basic properties of the mapping from genotype to phenotype are encoded in the DNA, therefore, the genotypes of the two organisms in this example cannot be completely equal.

The theoretical foundations of this chapter have first been published in 2009 [103]. The experimental results given in Sec. 5.3.1 have been presented there, too. The alternate ceGPM as well as the experimental results given in Sec. 5.3.3 have roughly been presented in a 2010 short paper [104], however, the model and method descriptions as well as the discussion of the results are substantially extended here. The third part of the experimental study (Sec. 5.3.6) has not been published before.

5.1 Overview of the Presented Approach

In this chapter, a method inspired by the evolvable GPM in nature is proposed to generalize evolutionary setups in artificial evolutionary systems in order to make the process flexibly adaptable to the circumstances in which they operate. The achieved flexibility is sufficient to allow for an adaptation of the effects of search operations, i. e., mutation and recombination, to the evolutionary search space during a run. This process can to some extent be compared to the capability of the well-known Evolution Strategies to evolve the mutation probability along with the phenotypes [13] (although Evolution Strategies provide a completely genotype-based approach). However, Evolution Strategies are only applicable when instances of the problem to solve can be encoded as a sequence of real values. While this is principally the case for most problems, in practice it is often hard to establish an adequate smoothness-preserving encoding (i. e., "small" mutations leading to "small" changes on the phenotype level). Particularly in ER, where genotypes tend to encode complex structures such as robot controllers rather than just parameters of a computational problem, it is often unclear how to generate a reasonable encoding for a specific scenario.

Going beyond the idea of Evolution Strategies, the approach presented here is capable of searching automatically for an adequate GPM from an integer-valued genotype into a complex phenotype during an evolutionary run. Using this type of flexible GPM, the effects of the search operations on the phenotypes can be altered as the interpretation of the genotypes changes during a run. Furthermore, in a recursive process, the effects of the search operations applied to the GPM itself can change at the same time, thus, potentially improving the evolution of GPMs as well as the evolution of controllers.

Therefore, in a typical evolutionary run, the genotype consists of two parts encoding a robot controller and a GPM, respectively, and search operations are applied to both parts. The GPM is called a *completely evolvable GPM (ceGPM)* here, indicating that not only the effects of search operations on the phenotype level are adaptable, but also the effects of search operations performed to the GPM-encoding part of the genotype on a GPM-altering level, and that this property does not require an additional explicit mapping from GPM encodings to GPMs. To accomplish this, both the genotypic and phenotypic representations of GPMs have to be structurally identical to those of robot controllers. As a result, the same search operators can be used for both types of genotypes, and the same GPM can be used to translate from integer sequences to robot controllers as well as to other GPMs. Due to the analogy of the translation processes, both the space of robot controllers and the space of GPMs are referred to as phenotypic spaces, cf. Sec. 5.2.4. To emphasize the active role of GPMs in translating genotypes into phenotypes, a GPM is alternatively called *translator* in the following.

In this chapter, FSMs are used to establish the algorithmic description of robot controllers and GPMs. The controller representation is given by the MARB model introduced in Chap. 4. An equivalent model represents the GPM, differing from the MARB model only in the commands produced and the sensory information perceived by the automaton. However, the idea can be adapted to other structures such as ANNs, learning classifier systems etc. as well. The only prerequirement is that the problem instances (robot controllers) and the GPMs are encoded by the same computational model, and that this model is computationally sufficient to provide for a "universal translator", i. e., a specific GPM that can produce any problem instance as well as any GPM – including itself – depending only on the genotype to translate, cf. Sec. 5.2.4.

The next section introduces the FSM-based approach as well as a set of evolutionary operators more formally. Furthermore, it gives definitions of basic terms and a detailed description of the evolutionary process.

5.2 A Completely Evolvable Genotype-Phenotype Mapping

In this section, an evolutionary approach is proposed which allows for the automatic improvement of the GPM at evolution time. A fundamental premise of the presented model is to use the same representation for both robot controllers and the GPM to make it possible to evolve both simultaneously using the same set of evolutionary operators. As before, the MARB model is used for the representation of robot controllers, and a structurally identical model called *Moore Automaton for Protected Translation (MAPT)* is used for the representation of GPMs. In the course of evolution, both are subject to mutation and can be evolved together.

5.2.1 Definition of (complete) evolvability

Since the work of Richard Dawkins [33], evolvability has been in the focus of research not only in evolutionary biology, but also in computer science leading to a new theoretical understanding of the term [43, 196]. Evolvability for artificial systems has been defined in multiple ways which particularly differ in the requirement of dynamics of the GPM. Wagner and Altenberg define it in a rather conservative, i. e., static way to be "the ability of evolutionary operators to sometimes produce improvements" [197]. Reisinger et al. define it more dynamically as "an adaptive organization of the genotype-phenotype mapping such that the search operators can produce more favorable phenotypes" [164]. The latter captures more accurately the biological view where evolvability is often thought of as "a character that can itself evolve" [181]. As the GPM plays an important role in establishing this capability, the definition used here in the context of ER is a more specific combination of the latter two:

Definition 5.1 (*Evolvability*)

> Evolvability *is defined to be the capability of an evolutionary system to coevolve the GPM along with phenotypic traits, in order to produce improvements in a variety of different evolutionary contexts by adapting the effects of the search operators (mutation and recombination). If, in a recursive manner, the same procedure is applied to the evolving GPM itself, such that effects of search operations performed on the GPM can be adapted, too, without requiring another static "GPM of the GPM", the system is called* completely evolvable.

A completely evolvable system, thus, allows for arbitrary search operators to evolve (with respect to the expressive power of the GPM) without predefining or preferring any particular method. As in nature, from a high-level perspective any evolutionary setup is evolvable, because the underlying static evolutionary setup which involves the ceGPM is flexible enough to allow for such changes in the operators. While parts of the GPM have successfully been evolved in the past, evolution of the complete GPM has been suggested in the literature [92], but to the author's best knowledge not suc-

Genotype

Behavioral gene

| 120 | 056 | 133 | 000 | ... | |

GPM/translator gene

| 004 | 188 | 251 | 012 | ... | |

GPM

Output
(phenotypes)

GPM

MARB
controller

MAPT
GPM/translator

Behavioral phenotype

behavioral

GPM

GPM/translator phenotype

Input (genes)

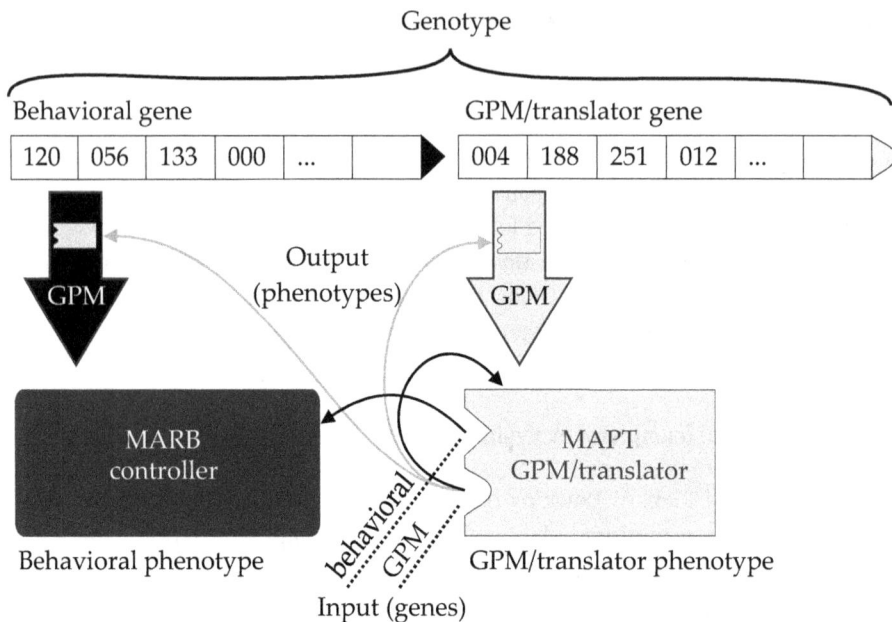

Fig. 5.2. Complete evolvability. Dividing the genotype into a behavioral and a GPM (or translator) part requires, in principle, two GPMs (big arrows labeled GPM) to produce the according phenotypic structures. The completely evolvable approach proposed here uses the same phenotype structure for both controllers (MARB) and GPMs (MAPT). Therefore, the GPM translated from the GPM gene can be used for translation of both the behavioral and, recursively, of the (future) GPM gene – making both GPMs evolvable while not requiring another non-evolvable level. (In the figure, the big MAPT box represents the same automaton as used by and suggested in the big GPM arrows.) Conversely, in a non-complete approach, an evolvable GPM would require another level involving a meta-GPM to translate the GPM gene, thus, inherently implying a non-evolvable sub-part of the system.

cessfully performed before. Fig. 5.2 shows an illustration of complete evolvability as opposed to a non-complete approach involving an evolvable GPM.

It is, however, important to assure that the descriptive model underlying the GPM has sufficient expressive power to allow for a variety of different search operators to evolve (otherwise complete evolvability can be a trivial property). A GPM/translator is called *fully evolvable* relative to a set of translators $\mathcal{P}^{trans}$ if in the course of evolution the alterations of the GPM can lead to any possible future interpretation of the genotypes representable by $\mathcal{P}^{trans}$. In other words, any mapping from the space of genotypes to the space of phenotypes can be evolved if it corresponds to a translator from $\mathcal{P}^{trans}$. The two ceGPMs introduced in this chapter are fully evolvable relative to the according translator model $MAPT$ or $MAPT'$, respectively, cf. Sec. 5.2.3.

5.2.2 Properties of ceGPM-based genotypic encodings

According to Nolfi and Floreano in their fundamental textbook on ER [142], genotypic encodings are supposed to have *expressive power* (i. e., many different phenotypic characteristics should be encodable), *compactness* (i. e., the length of the genotype should not directly reflect the complexity of the phenotype, e. g., by introducing loop structures), and *evolvability* (which is there defined as a general capability of the evolutionary operators to produce improvements), cf. Sec. 2.1.6. In the following, the applicability of these properties to a ceGPM approach based on an FSM controller structure is discussed.

1. Expressive power: In the presented approach, the whole GPM is part of the genotype of a robot. While this does not change the principle expressive power of MARBs encodable in the genotype, it does introduce another dimension of expressibility as the method of how individuals are mutated and recombined during evolution can be indirectly encoded in the genotypes. More precisely, a genotype encodes the way of interpreting the behavioral part as well as the translation part of a genotype. This new dimension of expressibility does not allow for completely new controller types that would not have been representable without a ceGPM. But it provides the genotype with the expressive power to encode a search strategy which can be adapted during evolution to more quickly find an optimum in a given search space.

2. Compactness: The translation from genotypes to phenotypes is performed by FSMs which can have loops, thus being able to create repeating structures and allowing for compact genotypic representations in the above-described sense. However, using other, more complex computational models such as ANNs to perform the GPM may further improve compactness.

3. Evolvability: As stated before, the evolvability of an evolutionary system is a multiply defined term and a property which is hard to identify. With respect to the definition of Nolfi and Floreano, the evolutionary approach defined below has been set up according to their suggestions to improve the system's ability to create improvements. However, as the GPM changes during evolution, this ability itself is expected to be further improved at runtime. As such dynamic properties are not captured by this definition of evolvability, experiments are performed to show the system's evolvability according to Def. 5.1. In Sec. 5.3.3 the results of these experiments are discussed finding evolvability to be present in one of the two ceGPMs proposed below.

Overall, the properties stipulated by Nolfi and Floreano are not directly applicable to ceGPM-based approaches. However, when generalized by preserving their intended meanings as suggested above, ceGPMs accomplish enhanced expressiveness and evolvability while not influencing compactness.

5.2.3 The Translator Model MAPT and the Course of Evolution

In this section, the controller and translator models as well as the evolutionary model and their connection to each other are described. There, the MARB model from Chap. 4 is briefly revisited and the translator model MAPT is introduced.

Overview of the two automaton types. As stated before, the MAPT model, which represents descriptions of translation processes from genotypes to phenotypes, is structurally identical to the MARB model, which represents robot controllers. For both types of Moore automata, each state of the automaton provides an output which is interpreted as an *operation* to be executed. This is a movement operation in the MARB case, and an "automaton construction" operation in the MAPT case, i. e., it describes a part of the construction process of a new (behavioral or translator) automaton, see below.

The transition function, which determines how the automaton's states are traversed, is defined based on values from *real* or *virtual* sensors. In the case of MARBs, the accessible sensors are seven real physical IR sensors which are placed on a robot and indicate distances to objects (cf. Figs. 4.6b and 4.7b in Chap. 4). In the case of translator automata, i. e., MAPTs, there are five virtual sensors providing data from the genotypic sequence to be translated and data from register fields; the latter can be used by a translator for temporary storage of byte values during the translation process.

Similar to classic Moore machines, a genotypic sequence serves to a MAPT as an input tape which is traversed from left to right using a *reading head*. Two of the five virtual sensors perceive the byte value at the reading head position (h_{100}) and the position right to it (h_{99}), respectively, the other three (h_{101}, h_{102}, h_{103}) serve as registers, each storing one byte value. Therefore, the only difference between MARBs (navigating a robot through an environment) and MAPTs (producing new automata by traversing a genetic sequence) are the different operation and sensor spaces they work on, cf. Fig. 5.5 for a depiction of the translation process. Fig. 5.3 shows an example MARB and an example MAPT, both non-evolved, but hand-crafted.

Overview of the evolutionary system. Every robot carries its own MAPT translator and, as before, an individual MARB controller which both can be altered during evolution by applying the same evolutionary operators to their genotypes. At the beginning of a run, the initial MARB is empty, and the initial MAPT is set to the hand-crafted translator depicted in Fig. 5.3b which has a specific universality property (cf. Sec. 5.2.4). Every robot has a two-part genotype belonging to the two automaton types, called *behavioral gene* and *translator gene*, respectively, both being sequences of byte values. During evolution, mutation is applied to both genes. After a mutation occurred, the respective mutated gene is translated by the robot's MAPT into another MARB or MAPT, respectively, which replaces the old MARB or MAPT of the robot. There, the replacement of the old MAPT by a new mutated version of itself establishes

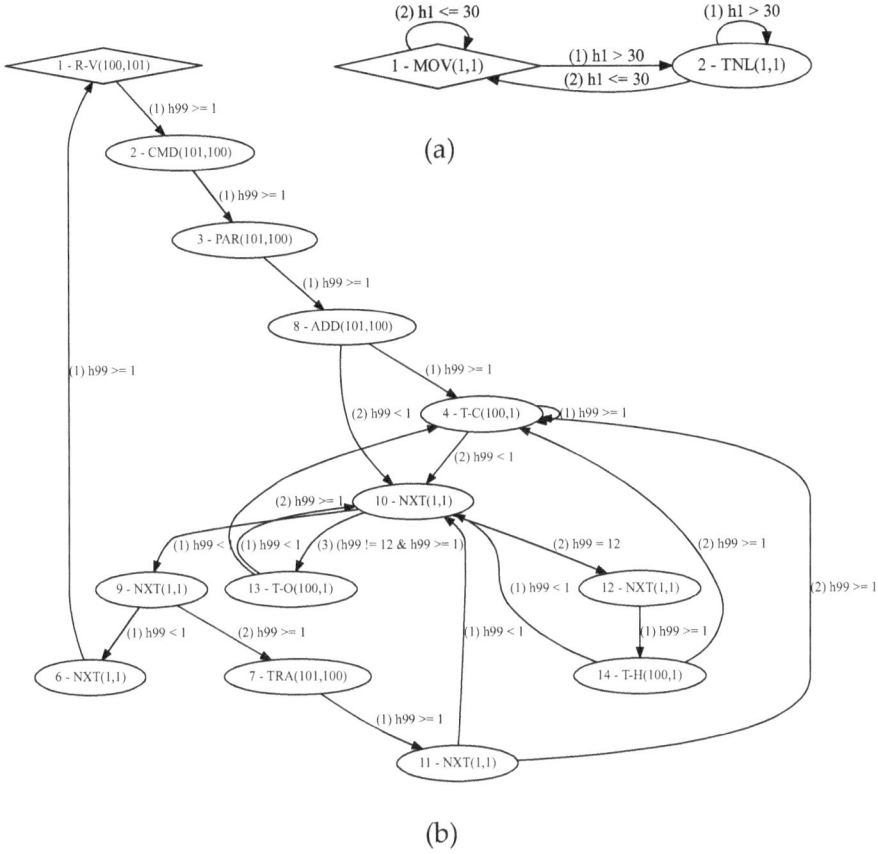

Fig. 5.3. Two hand-crafted example automata. (a) Example MARB performing a simple CA behavior (drive forward: (*Move*, 1, 1) as long as no obstacle is ahead: $h_1 \leq 30$; turn left: (*TurnLeft*, 1, 1) if an obstacle is ahead: $h_1 > 30$). (b) Example MAPT, built to perform a "universal translation" meaning that any MARB or MAPT can be generated if an according genotypic sequence is used (cf. Sec. 5.2.4). This MAPT is used as initial translator of the evolutionary runs in the first part of the study described in Sec. 5.3. (Note that the commands T–H, T–C, T–O, R–V, ... are written as *TH*, *TC*, *TO*, *RV*, ... in the subsequent text, omitting the "–" character for better readability.)

the recursive process which makes the system completely evolvable. By this means, both the MARB and MAPT automata can evolve simultaneously without requiring an explicit underlying fixed GPM.

In contrast, similar approaches which coevolve the search operators usually require a fixed top-level set of operations that cannot be altered. Particularly, in Evolution Strategies the probability of mutations to occur can be altered, but the actual mutation operation including its effects on the phenotype level remains unchanged. Other approaches provide a set of operations to mutate mutation, but they cannot mutate the mutation of mutation, unless another top-level mutation is introduced. Here, by using a recursive "self-translation" of the GPM, there is no need for a fixed top-level GPM to establish the mutation of a GPM. When the GPM is mutated, it changes the way future mutations on the MARB level as well as on the MAPT level, i. e., the GPM itself, are interpreted, allowing for a complete adaptation of the search operators to the environmental circumstances.

As before, selection is performed according to a fitness value which is calculated by observing the behavior of a robot in the environment, i. e., solely based on a robot's MARB controller. Fig. 5.4 depicts an overview of the evolutionary process (cf. also Fig. 5.2).

Preliminaries. To keep notations consistent with the previous chapter, symbols concerning MARBs will be used without indices while symbols concerning a part X of a MAPT will be assigned the superscript index "*trans*": X^{trans}. For the purpose of a simpler notation, "$\star$" is used as a "do not care" symbol which can be empty or *trans*.

A (real or virtual) sensor is represented by a sensor variable yielding a byte value. As before, $B =_{def} \{0, \ldots, 255\}$ and $B_+ =_{def} \{1, \ldots, 255\}$ denote the sets of all and only the positive byte values, respectively. A sensor variable storing the sensor value $v_i(t)$ at a time step t is denoted by h_i. For $i \leq j \in \mathbb{N}$ and a set of sensor variables $\{h_i, h_{i+1}, \ldots, h_j\}$, the tuple $(v_i(t), v_{i+1}(t), \ldots, v_j(t)) \in B^{j-i+1}$ denotes the actual corresponding values of the sensors at a time step t; in the following, the notation of time step t can be omitted for the purpose of better readability.

In the case of behavioral automata, the set $H =_{def} \{h_1, \ldots, h_7\}$ defines the sensor variables corresponding to the seven IR sensors of a Jasmine IIIp robot. For $1 \leq i \leq 7$, h_i is associated to the real sensor labeled with i in Fig. 4.6b of Chap. 4. At a specific time step t in an environment, the tuple of current values of the sensors is $v(t) =_{def} (v_1(t), \ldots, v_7(t)) \in V = B^7$.

In the case of translator automata, the sensor variables are defined by the set $H^{trans} =_{def} \{h_{99}, \ldots, h_{103}\}$. For $99 \leq j \leq 103$, h_j is associated to the virtual sensor labeled with h_j in Fig. 5.5. At a specific point in time $t' \in \mathbb{N}$ during the traversal of a genotypic sequence (t' is given by the number of state changes performed by the automaton), the tuple of current values of the sensors is $v^{trans}(t') = (v_{99}(t'), \ldots, v_{103}(t')) \in V^{trans} =_{def} B^5$. There, $v_{100}(t')$ represents the value at the current position of the reading head and $v_{99}(t')$ the value right to the position of the reading head or zero if the

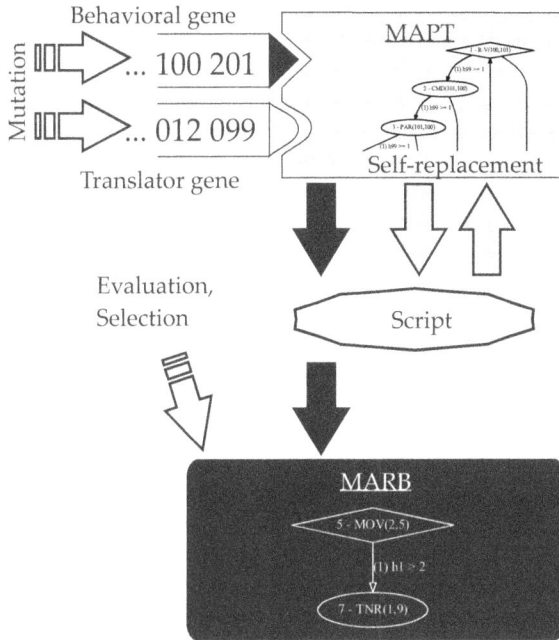

Fig. 5.4. Schematic view of the mutation process onboard of a robot. As a purely genotypic operation, mutation affects the behavioral or the translator gene sequence only, by adding random
(Gaussian-distributed) values to randomly chosen numbers in the respective sequence. After a
mutation of one of the genes, a robot's current MAPT uses the altered sequence as an input and
translates it to a new MARB or MAPT (depending on which sequence has been mutated), by which
the former MARB or MAPT is replaced, respectively. Particularly, to establish complete evolvability,
the MAPT replaces itself after a translator gene mutation by a translation of the mutated translator
gene. As an intermediate step in the translation process, the MAPT first generates a script, which is
executed to create the actual automaton.

reading head is at the rightmost position. The reading head starts at the leftmost position at $t' = 0$ and moves one position to the right at each time step. The values
$v_{101}(t'), \ldots, v_{103}(t')$ represent three registers which can be fed with byte values to
store information about the previous course of translation. Note that the notion of time
during the translation process has a different dimension than environmental time. In
simulation, environmental time is paused during translation , i. e., the whole translation process takes place between two environmental time steps t and $t + 1$; on real
robots the translation process is expected to require a negligibly short amount of time.

Only one of the sets of sensor variables H or H^{trans} is used at the same time, depending on the usage in the context of MARBs or MAPTs. The omitted sensor variables
$h_8, \ldots, h_{98}$ are not used in this thesis. The gap can be used to introduce additional
sensors without loosing backwards compatibility.

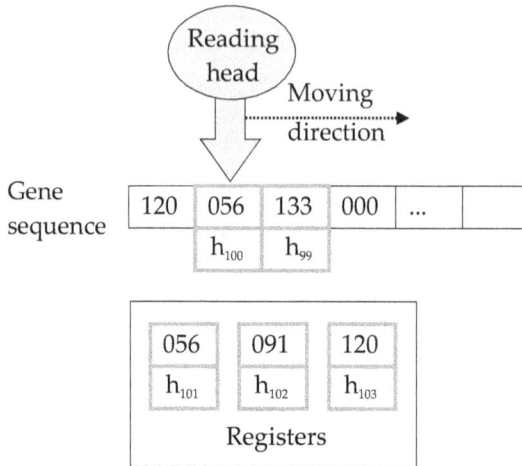

Fig. 5.5. Virtual sensors of a MAPT during a translation process. The figure schematically shows the translation process traversing a (behavioral or translator) gene sequence with a unidirectionally moving reading head (random byte values are depicted as examples). Sensors h_{100} and h_{99} provide the reading head data by returning the byte value given at the current position of the reading head and a lookahead value one position right to it, respectively (h_{99} is set to zero if the reading head is at the rightmost position). The sensors h_{101}, h_{102} and h_{103} serve as registers storing one byte value each of which can be set by a MAPT using the RC and RV operations (see below). These five virtual sensors determine a MAPT's active transitions just as the real IR sensors do in the MARB model, cf. Chap. 4.

As before, a randomized function $rand$ is assumed which returns a random element out of an arbitrary set based on uniform distribution; furthermore, $rand_{gaussian}$ is a function that returns a random real number by standard normal distribution (with a mean of 0 and a standard deviation of 1). In reality, depending on the implementation of the corresponding random number generator, both these functions may vary in their behavior. Particularly, $rand_{gaussian}$ is assumed to return a floating point decimal with double precision rather than an actual real value. For the experiments described here, the Java built-in random number generator as implemented within the EAS Framework is used for the generation of all random numbers.

Automaton Definition. The following description is based on the already defined MARB model which remains unchanged for the representation of robot behavior. The MAPT model is structurally equivalent, differing only in the sets of sensor variables and operations. Both MARBs and MAPTs are Moore Automata

$$(Q, \Sigma, \Omega, \delta, \lambda, q_0)$$

where Q is the set of states, Σ is an input alphabet, Ω is an output alphabet, $\delta : Q \times \Sigma \to Q$ is a transition function, $\lambda : Q \to \Omega$ is an output function, and $q_0 \in Q$ is the initial state. The execution begins at the initial state q_0; for every state $q \in Q$, an output $\lambda(q)$ is given, which is here interpreted by executing an operation; a unique following state $\delta(q, e)$ is determined by the transition function based on a symbol $e \in \Sigma$ which is read from the input stream. In the case of MARBs, the input stream is given by the seven sensors of a robot representing its perceptions while it moves through an environment, for MAPTs the input stream is given by the five virtual sensors resulting from the traversal of a genetic sequence from left to right, while performing a state change at every time step. The output at each state defines a movement operation for MARBs and a construction operation for MAPTs. The available operations for MARBs and MAPTs are given below. For both MARBs and MAPTs, the flexible parts of the automata which are subject to evolution are:

- the set of states Q including q_0,
- the output function λ, and
- the transition function δ.

The output alphabet is fixed and given by a set of operations $\Omega = Op^*$ as defined below for behavioral or translator automata, respectively. The input alphabet is fixed, too, and consists of all possible combinations of sensor values $\Sigma = V^*$ for behavioral or translator automata, respectively. As before, a state q is identified by a positive byte value id^q. The space of all MARBs is called $MARB$, the space of all MAPTs is called $MAPT$ in the following.

Conditions. For MARBs, the set C of conditions over the sensor variables H has been defined by (cf. page 72):

$$c ::= true \mid false \mid z_1 \triangleleft z_2 \mid (c \circ c),$$

where $z_1, z_2 \in B_+ \cup H$,
$\triangleleft \in \{<, >, \leq, \geq, =, \neq, \approx, \not\approx\}$,
$\circ \in \{AND, OR\}$.

For translator automata, the condition set C^{trans} is defined in the same way, except for using the set of sensor variables H^{trans} instead of H. The evaluation function S, as defined before, is extended in the obvious way to fit both behavior and translator automata:

$$S : C^* \times V^* \rightarrow \{true, false\}$$

It evaluates a condition at a time step t to $true$ or $false$ by assigning the current sensor values $v(t) \in V^*$ to the sensor variables from H^* which C^* is based on.

As before, two cases of inconsistencies can occur and have to be considered for both automaton types. (1) If for a state none of the outgoing transitions has a condition that evaluates to $true$ (*active* condition), there is an implicit default transition to the initial state. For example, both transitions pointing to the initial state 1 in Fig. 5.3 (a) could be deleted without causing a change to the automaton behavior as they are implicitly defined. (2) If more than one condition evaluates to *true*, the first of the *true* transitions is chosen, based on the order of insertion during construction of the automaton.

Operations. The operations have the form of a tuple $(cmd, par_1, par_2) \in B_+ \times B_+ \times B_+$ of byte values, consisting of a command cmd and a first and second parameter par_1 and par_2. The command cmd is from one of the sets $Cmd \subseteq B_+$ or $Cmd^{trans} \subseteq B_+$. The parameters may be ignored for some commands; particularly, in the MARB case the second parameter is ignored for all commands. In the following, an operation (cmd, par_1, par_2) is alternatively written in function notation as $cmd(par_1, par_2)$, $cmd(par_1)$ or $cmd()$, if no parameter, the first parameter or both are ignored, respectively.

Operations for MARBs. As before, behavioral commands are defined by the set

$$Cmd = \{Move, TurnLeft, TurnRight, Stop, Idle\}$$

which is internally mapped on byte values, i. e., $Cmd \subset B_+$. The set of all possible behavioral operations is defined by

$$Op = Cmd \times B_+ \times B_+$$

with the meaning for an operation $cmd(par_1)$ (as the second parameter is ignored for all behavioral commands): drive forward for at most par_1 mm, turn left for at most par_1 degrees, turn right for at most par_1 degrees, stop performing the current operation, or keep performing the current operation, respectively. For *Stop* and *Idle*, the parameter par_1 is ignored. An example for a behavioral automaton's operation is $Move(10) \in Op$.

Operations for MAPTs. A translator's purpose is to define all the flexible parts of an automaton, namely states, transitions, and the according state outputs and conditions, depending on a gene to be translated. The language of conditions, however, is context-free and cannot be expressed by a pure finite state model. Moreover, it is desired that every operation executed by a translator has a well-defined and strictly determined effect on the outcoming automaton and that this effect is "non-trivial" for "most" operations. E. g., if an operation for inserting a transition between two states a and b would be defined to work only if the states a and b already exist in the automaton, this operation would have no effect for most combinations of a and b in most automata of usual size due to the randomized creation of genes. Similarly, if the translation of a condition would be mapped on a trivial condition (e. g., *true*) any time a syntactically incorrect part occurs, practically all conditions would be translated to *true*. (The terms "non-trivial" and "most" are used rather intuitively here, as in the above examples. Overall, trivial effects of an operation involve (1) no effect at all, (2) oscillating effects that erase each other and (3) effects producing a constant automaton (part) without considering the gene to translate). While such effects are rather

harmless if they occur occasionally, they can corrupt the idea of evolvable translators
if they occur too frequently.

To both solve the problem of context-free conditions and avoid trivial effects during translation, a script language has been introduced which provides access to a stack
memory for the construction of conditions. As part of the interpretation of this script
language, a repair mechanism is performed on invalid operations to make their effects
valid, and, in most cases, non-trivial. A translator only has to create a valid script out
of a gene sequence while correctness, well-definedness and non-triviality are assured
during execution of the script. The name *Moore Automaton for Protected Translation*
is dedicated to this repair mechanism which protects the GPM from syntactical failures and triviality during translation. (Additionally, when pronounced, MAPT should
sound like "mapped".)

In the following, the script language for MAPTs is defined. A script is a sequence
of operations, which build a subset of the actual MAPT operations Op^{trans} defined
below. A *script* of length $n \in \mathbb{N}$ is a sequence $scr \in (Op^{scr})^n$ of script operations
$cmd(par_1, par_2)$ to be executed sequentially. There, the set of script operations is defined as

$$Op^{scr} =_{def} Cmd^{scr} \times B \times B \subset Op^{trans}$$

with

$$Cmd^{scr} =_{def} \{nod, cmd, par, add, TH, TC, TO, edg\} \subset Cmd^{trans}.$$

As in the MARB case, the commands are internally mapped on byte values, i.e.,
$Cmd^{scr} \subset Cmd^{trans} \subset B_+$. The commands $nod, cmd, par,$ and add are used to insert
a state, to change its command, or its first or second parameter, respectively. TH,
TC, and TO are commands for building a condition in the internal stack memory by
inserting sensor variables (TH), atomic constants (TC) and operators (TO) in a postfix manner. This is the most complex part of the translation process as a syntactical
correctness of the final condition has to be guaranteed by simultaneously assuring
non-trivial translation. It is achieved by changing the type of a postfix element if it
does not syntactically fit at its position.

Only one condition at a time can be built up during translation meaning that a
condition has to be completed before another one can be generated. The command
edg inserts a transition between two states using the currently built-up condition. If
edg is invoked while the current condition is not yet finished, the condition is finished by using a deterministic (non-trivial but reproducible, see below) procedure of
inserting postfix elements, before the condition is inserted into the automaton. If the
parameter of an operation is out of range for that operation, the script interpreter maps
it on the adequate range via a modulo operation (this procedure is assumed to be done
implicitly in the following). Invalid values are replaced using a *standard value generator*; it generates standard values as values of specific counters for each of the different value spaces; a counter is increased every time a value from its value space is
used. Particularly, there are counters for the sets of sensor variables, atomic constants

Behavioral
gene

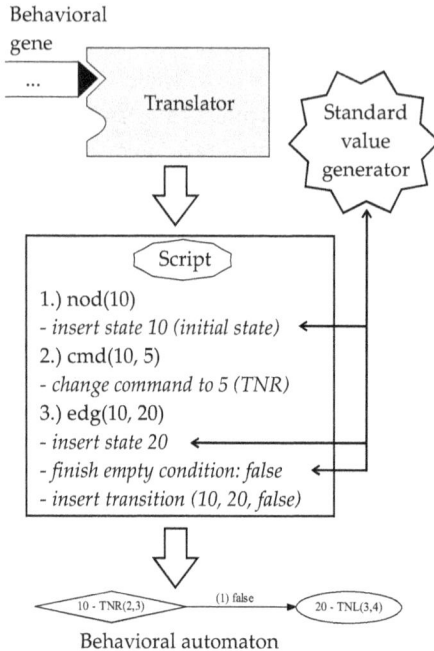

Behavioral automaton

Fig. 5.6. Example translation of a behavioral gene into a MARB. As a first step, a MAPT generates a script out of a behavioral gene. The script operations are interpreted sequentially performing a repair mechanism on invalid operations using a standard value generator. In the figure, the *edg* operation is executed without any condition having been created before. Therefore, the repair mechanism creates the constant *false* as according condition (*true* would have been possible too, depending on the state of the standard value generator). In the same way, a translator gene can be translated into a new translator.

and the two types of operators (comparison and conjunction/disjunction), which are used to provide a non-random uniform distribution of the elements of a condition, cf. description of operation $edg(X, Y)$ below. Fig. 5.6 shows an example for the translation process using a script. In detail, the script operations are interpreted as follows $(X, Y \in B_+$; a state q is identified by its name $id^q \in B_+)$:

- $nod(X)$: Insert state X (if state X already exists do nothing; if X is the first inserted state, declare it initial state), and use a standard command, parameter, and additional parameter to build the state's operation.
- $cmd(X, Y)$: If state X does not exist, execute $nod(X)$. Change the command of the state's operation to Y.
- $par(X, Y)$: If state X does not exist, execute $nod(X)$. Change the first parameter of the state's operation to Y.

- $add(X, Y)$: If state X does not exist, execute $nod(X)$. Change the second parameter of the state's operation to Y.
- $TH(X)$: Add X to the currently built-up condition in the stack as next postfix-symbol if a sensor variable is syntactically correct here, otherwise execute $TO(X)$.
- $TC(X)$: Add X to the currently built-up condition in the stack if an atomic constant ($true$ or $false$) is correct here, otherwise execute $TO(X)$.
- $TO(X)$: Add X to the currently built-up condition in the stack if a comparison operator ($<, >, \ldots$) or conjunction/disjunction operator (AND/OR), respectively, is correct here; otherwise consult the standard value generator and execute $TH(X)$ or $TC(X)$ accordingly. (Note that one of the commands TH, TC or TO has to lead to a syntactically correct extension of the built-up condition, and thus the recursion has a valid breaking condition.)
- $edg(X, Y)$: If state X does not exist execute $nod(X)$; if state Y does not exist execute $nod(Y)$. If the current condition $cond$ is unfinished, finish $cond$ by using the standard value generator. Insert a transition from state X to state Y labeled $cond$.

Note that this basic set of script operations and, thus, the set of translator operations is extended by the additional command $cpl(X, Y)$ for an improved version of the MAPT model in Sec. 5.3.3.

The set of MAPT commands is a superset of the script commands and defined to be

$$Cmd^{trans} =_{def} Cmd^{scr} \cup \{RC, RV, NXT\}.$$

The set of MAPT operations is accordingly defined to be

$$Op^{trans} =_{def} Cmd^{trans} \times B_+ \times B_+.$$

If the output of a MAPT state is a script operation $A(X, Y) \in Op^{scr}$, this has the meaning of inserting $A(X, Y)$ at the end of the current script. The new non-script commands are two register commands RC, RV, and an idle command NXT. The according operation $NXT()$ does not perform any action, the automaton only moves on to the next state while the reading head moves one step further to the right. The register operations $RC(X, Y)$ and $RV(X, Y)$ store in the register corresponding to sensor variable h_X

- the *constant* value Y or
- the *variable* value from sensor variable h_Y, respectively.

If X or Y are out of range, a modulo transformation is performed to adjust them to the range of register sensors in the case of X, i. e., $h_{101}, \ldots, h_{103}$, and to the range of all sensors in the case of Y in combination with command RV, i. e., $h_{99}, \ldots, h_{103}$. For a formalization of the translation process, cf. Alg. 5.1 at the end of the next section. Furthermore, the actual Java implementation of the translation process can be inspected in package `eas.simulation.spatial.sim2D.marbSimulation.translator` of the EAS Framework [101].

The MAPT model is capable of generating any possible automaton from the spaces of all MARBs or MAPTs. More precisely, for each (behavioral or translator) automaton A there exists at least one combination of translator t and gene g, such that the translation of g by t yields A, i.e., $dec^*(t, g) = A$ in the notation introduced in Sec. 5.2.4. It even holds that independently of the sequence g there exists a translator t with $dec^*(t, g) = A$, as long as the sequence g is sufficiently long, meaning that there are MAPTs which completely ignore the sequence g during translation. The more important inverse property (i.e., any MARB or MAPT being producible by a single MAPT, depending only on the sequence g) holds for the sub-class of universal translators only, see below.

5.2.4 Genotypic and Phenotypic Spaces

In this section, the genotypic space and two phenotypic spaces, i.e., a space of robot behaviors and a space of translators, are defined. The genotypic space consists in both the MARB and the MAPT case of sequences of byte values which are mapped on behavioral or translator automata in the respective phenotypic spaces.

Genotypic space. A *gene* or *genetic sequence* is a flexible-size array of byte values from the *genotypic space*

$$\mathcal{G} =_{def} \bigcup_{n \in \mathbb{N}} B^n.$$

In the following, the byte value at position i in a gene $g \in \mathcal{G}$ is denoted by $g[i]$, and $|g|$ denotes the length of g, i.e., the number n of byte values in the array. A genome G^R of a robot R consists of two genes, namely the behavioral gene g^R and the translator gene $g^{R,trans}$, and is given by the tuple

$$G^R = (g^R, g^{R,trans}) \in \mathcal{G} \times \mathcal{G}.$$

This notion is preferred over the alternate possibility of defining $\mathcal{G} \times \mathcal{G}$ as a single genotypic space, as the genotypic spaces of behavioral and translator automata are treated separately for the most parts of the model.

Phenotypic spaces. Two different phenotypic spaces are distinguished:
1. The space of (behavioral) phenotypes $\mathcal{P} =_{def} MARB$ is the space of all robot controllers. Its elements are, in a classic terminology, the actual phenotypes which are executed in an environment and which are directly subject to evaluation and selection.
2. The space of translator phenotypes $\mathcal{P}^{trans} =_{def} MAPT$ is the space of all translators or evolvable GPMs. Translators from $\mathcal{P}^{trans}$ translate behavioral genes and translator genes into robot controllers (MARBs) and other translators (MAPTs), respectively. They are evaluated only indirectly as they influence evolvability during a run causing better or worse long-term performance (there is one exception to this rule as empty translators are sorted out directly after mutation; cf. Sec. 5.2.5).

There exist two mappings (*decodings*) from the genotypic space to the two phenotypic spaces:

$$dec : \mathcal{G} \times \mathcal{P}^{trans} \rightarrow \mathcal{P}, \tag{1}$$

$$dec^{trans} : \mathcal{G} \times \mathcal{P}^{trans} \rightarrow \mathcal{P}^{trans}. \tag{2}$$

The mapping *dec* denotes the creation of a robot behavior out of a gene using an existing translator. The mapping dec^{trans} denotes the creation of a new translator out of a gene using an existing translator. Therefore, both decodings do not only depend on the gene which is being decoded, but also on the translator, and different translators can produce different outcomes for the same gene.

Partial completeness of $\mathcal{P}^{trans}$. The former statement can be further generalized as it holds that any behavioral (or translator) automaton can be produced by *dec* (or dec^{trans}) out of any gene g depending only on the decoding translator, and the length of g (there, $|g|$ has to exceed a value k depending on the output automaton, as a translator performs exactly $|g|$ state changes during a translation):

$$\forall g \in \mathcal{G}, \forall b \in \mathcal{P} \, \exists t \in \mathcal{P}^{trans} \, \exists k \in \mathbb{N} : |g| \geq k \Rightarrow dec(g, t) = b, \tag{1}$$

$$\forall g \in \mathcal{G}, \forall t' \in \mathcal{P}^{trans} \, \exists t \in \mathcal{P}^{trans} \, \exists k \in \mathbb{N} : |g| \geq k \Rightarrow dec^{trans}(g, t) = t'. \tag{2}$$

This can be proven easily by constructing for each behavioral automaton b (or translator automaton t') a translator t which ignores the input and creates as a constant output the MARB b (or the MAPT t'). This property is called the *partial completeness* of the space of translator phenotypes $\mathcal{P}^{trans}$.

Universal translators. A direct implication of the above completeness statement is that for a fixed translator there may exist behavioral or translator automata which it cannot create, regardless of which gene is being decoded (e. g., for the aforementioned constant translators, most behaviors or translators cannot be created as these translators always produce the same output). Translators, therefore, have a higher impact on the outcome of translations than the genes to be translated.

There is, however, a class of translators, called *universal translators*, which can produce every possible behavioral or translator automaton depending on the gene to be translated. A universal translator is a translator u which meets one of the following requirements:

$$\forall b \in \mathcal{P} \, \exists g \in \mathcal{G} : dec(g, u) = b, \tag{1}$$

$$\forall t' \in \mathcal{P}^{trans} \, \exists g \in \mathcal{G} : dec^{trans}(g, u) = t'. \tag{2}$$

In fact, a universal translator meeting requirement (1) can be transferred into a universal translator meeting requirement (2) and vice versa by just changing the range of output values produced by the translator. Therefore, in the implementation of the model, only one universal translator is used which is switched between the two ranges to translate both behavioral and translator genotypes. Alg. 5.1 shows, how a single translator can be used in two different modes by adjusting the sensor and operation ranges for the translator's outputs.

An important consequence of the above statements is that for a universal translator u meeting condition (2), there exists a "self-reflexive" gene $g_u \in \mathcal{G}$ such that the translation of that gene by the universal translator is the universal translator itself, i. e., $dec^{trans}(g_u, u) = u$. As in the beginning of an evolutionary run all parts of the phenotypic spaces should be reachable, a universal translator u is used as initial translator in all experiments. To allow for a continuous start, the initial translator gene is set to a self-reflexive gene g_u. Therefore, minor mutations on the translator gene are not expected to drastically modify the initial translator. During evolution, however, it is possible that non-universal translators occur which leads to not every behavioral or translator phenotype being reachable anymore – although always every genotype is reachable (cf. definition of the mutation operator in Sec. 5.2.5).

The two universal translators used in the experiments, cf. Figs. 5.3b and 5.13, both implement the encoding of MARBs (which can also serve as an encoding of MAPTs) presented in Sec. 4.1.3 of the previous chapter, for all syntactically correct genes in terms of the context-free grammar given there, too. For syntactically incorrect genes the introduced repair mechanism is performed during translation of the script leading to automata belonging to a preferably "close" syntactically correct gene.

The translation algorithm. Alg. 5.1 shows the translation process of a genotypic sequence $g \in \mathcal{G}$ by a MAPT $t \in \mathcal{P}^{trans}$ into a MARB or another MAPT a^*, depending on the translation mode. The translation mode is given by a Boolean flag beh which is *true* if the output is intended to be a behavioral automaton, and *false* for a translator automaton.

5.2.5 Evolutionary Operators

In this section, a new mutation operator $M_G : \mathcal{G} \rightarrow \mathcal{G}$ is introduced and a recombination operator is suggested which both work on the genotypic space $\mathcal{G}$, consisting of sequences of byte values (cf. Sec. 5.2.4). Furthermore, the selection and reproduction procedures as well as fitness calculation are described.

Mutation and temporarily unstable translators. A benefit of an evolvable GPM is that mutation and recombination do not have to be selected as carefully as with a static GPM. Rather, desired properties such as smoothness on the phenotypic level are expected to evolve during a run. Therefore, a standard mutation operator, well-studied for real-valued genotypes in classic EC, is adopted for M_G.

As in the experiments of Chap. 4, mutation is applied in constant intervals of length t_{mut} $\left(t_{mut}^{trans}\right)$ to a robot's behavioral (translator) gene. Since the experiments in this chapter have been performed solely in simulation, mutation is performed synchronously to all robots timed by a global clock. Where real-valued operations are

---

input : MAPT $t \in \mathcal{P}^{trans}$, genotypic sequence $g \in B^n$, $n \in \mathbb{N}$, Boolean value
$beh \in \mathbb{B}$ denoting if in behavior mode (or else in translator mode).
output: Automaton MARB or MAPT $a^* = dec^*(g, t) \in \mathcal{P}^*$.

set a^* to empty automaton and create empty *built-up-condition*;
// Note that the following line is the only one where *beh* is used.
if *beh* **then**
| set valid sensor variables to H and set valid operations to Op;
else
| set valid sensor variables to H^{trans} and set valid operations to Op^{trans};
end

$h_{101} := 0$, $h_{102} := 0$, $h_{103} := 0$, $(cmd, X, Y) :=$ output of t's initial state;
for int $i = 1$ to $|g|$ **do**
| $h_{100} := g[i]$, $h_{99} := g[i + 1]$;
| **if** $cmd =$ 'nod' **then**
| | insert state named h_X into a^*;
| **else if** $cmd =$ 'cmd' **then**
| | set command of a^*'s state h_X to h_Y;
| **else if** $cmd =$ 'par' **then**
| | set first parameter of a^*'s state h_X to h_Y;
| **else if** $cmd =$ 'add' **then**
| | set second ("additional") parameter of a^*'s state h_X to h_Y;
| **else if** $cmd =$ 'C_*' for $* \in \{C, H, O\}$ **then**
| | add h_X to *built-up-condition*;
| **else if** $cmd =$ 'edg' **then**
| | build condition *cond* by finishing *built-up-condition*;
| | insert transition from h_X to h_Y, associate *cond*;
| | empty *built-up-condition*;
| **else if** $cmd =$ 'R_C' **then**
| | $h_X := Y$;
| **else if** $cmd =$ 'R_V' **then**
| | $h_X := h_Y$;
| $(cmd, X, Y) :=$ output of t's next state;
end
return a^*

Algorithm 5.1: Translation of a genotype $g \in \mathcal{G}$ by a MAPT $t \in \mathcal{P}^{trans}$ into a MARB (behavioral flag $beh = true$) or a MAPT ($beh = false$) a^*. The flag beh is used only to determine the sensory and operation spaces. Whenever $g[i]$ is out of range for a certain i, it is assumed to be set $g[i] := 0$. The repair mechanism is assumed to be applied where necessary according to the description in Sec. 5.2.3.

used on the byte-valued genomes, a rounding operation ("round towards zero"), and a mapping to the range of byte values (by setting negative values to 0 and values above 255 to 255) is assumed to be performed afterward. M_G is given by applying the following procedure to a behavioral (translator) gene g: by a small probability ϵ $(\epsilon^{trans}) \in [0, 1]$, a normally distributed random value with a standard deviation of d $(d^{trans}) \in \mathbb{R}$ is added to every byte of gene g. In other words, the value $m_i \left(m_i^{trans}\right)$ is added to every byte $g[i]$ with:

$$m_i^* = \begin{cases} rand_{gaussian} \cdot d^* & \text{if } rand\,([0, 1]) \le \epsilon^*, \\ 0 & \text{otherwise} \end{cases}$$

Additionally, with probability $\epsilon^* \cdot f_+^*(|g|)$, a new byte $g[|g| + 1] =_{def} rand_{gaussian} \cdot d^*$ is appended to the end of the gene; with probability $\epsilon^* \cdot f_-^*(|g|)$, the rightmost byte is deleted. There, $f_+^*, f_-^* : \mathbb{N} \to \mathbb{R}$ are functions that provide for short genes a higher probability to get longer than for long genes, cf. definitions in Tab. 5.1.

After the mutation of a behavioral gene, the gene is translated by the current translator of the robot into a new MARB to let the mutation take effect. As well, after the mutation of a translator gene g, the current translator t translates the new gene $g_{mut} =_{def} M_G(g)$ and gets replaced by the new translator: $t \leftarrow t_{mut} =_{def}$ $dec^{trans} (g_{mut}, t)$, cf. Fig. 5.4. However, for translators this may lead to an unstable state, because a translation of g_{mut} by t_{mut} can again lead to a new translator different than t_{mut} and so on causing an unpredictably long chain of changes to the translator even when no further mutations affect the genotype.

To avoid unstable translators during evolution, after each translator mutation the replacement $t \leftarrow dec^{trans} (g_{mut}, t)$ is performed in a loop, as long as the resulting translator keeps changing. To deal with possible oscillations, a maximum number of 100 translations is not exceeded. Experience shows that this procedure usually converges quickly after few replacements and re-translations (virtually always in less than 10 iterations, except for the rare case of oscillations), and that the resulting translator is usually non-trivial. For this translator, the mutated gene g_{mut} is self-reflexive in the above-defined sense.

After mutation and translation to a stable translator, another small intrusion to the system is performed. In some cases, translators tend to degenerate to empty automata by mutation. As empty translators always produce empty output, i.e., non-moving MARBs and empty output producing MAPTs, it seems plausible to prohibit such mutations. In these cases the mutation operation is undone and the translator remains unchanged.

To provide comparisons to the experiments described in the previous chapter, experiments have also been made with the mutation operator M defined on page 86 of Chap. 4, using the best parameter setup found there as a benchmark.

Remarks on temporarily unstable translators. The problem of temporarily unstable translators comes from the core of the idea of ceGPMs. It lacks a counterpart in nature and has to be solved artificially which makes it (as any fixed human-built input to an evolutionary system) potentially fragile and a good starting point when looking for possible improvements of the process. The proposed method of re-translating to a stable state is a brute-force solution that appears to work satisfyingly well when using an FSM-based approach. However, with other translator types it may not be as easy to solve the problem. There, the chain of re-translated translators might fail to converge. Particularly, translators based on ANNs seem to lack this property, producing ever new translators by re-translation which neither oscillate between fixed instances nor seem to be similar in any relevant aspect. Another idea that might provide a solution is to make the translation function dec^{trans} invertible (regarding the invertibility of ANNs, cf. [72, 73, 90]). Then, for a given translator t, a new gene $g_t = (dec^{trans})^{-1}(t, t)$ can be calculated that has the desired self-reflexive property $dec^{trans}(g_t, t) = t$. This gene can replace the mutated gene and provide for a stable state of the translator. In contrast to traditional EA approaches, the consequence would be that the genotype would have to change apart from evolutionary operations, and without any changes on the phenotypic level.

Recombination. Focussing on the investigation of evolvability, for simplicity reasons no recombination operator is applied in the experiments described in this chapter. However, due to the elementary genotypic space $\mathcal{G}$, common recombination operators such as n-point crossover etc. can be easily adopted. Experimental studies including recombination operators are planned to be conducted in the future. Furthermore, it remains to be investigated to what extent the schema theorem [77] can be applied to the approach when using recombination. (Using the recombination operator from Chap. 4, the schema theorem cannot be applied in an obvious way.)

Reproduction and selection. Selection is based on the same tournament-based process as in the previous chapter. In a real-world version, robots produce offspring when they come spatially close to each other, in simulation the synchronized version is used which allows for more control of the selection process. There, all robots reproduce simultaneously according to a global clock, using the reproduction interval t_{rec}. At reproduction time, every robot builds a tournament with the $p-1$ robots closest to itself, with p being the number of parents, cf. Sec. 5.3. In a tournament, the robot with the highest fitness copies relevant parts of its onboard genotypes and phenotypes to the other robots. In runs with classic phenotype-level mutation, only the MARB phenotypes are exchanged during reproduction. In runs using the new genotype-level mutation operator M_G, but without the ceGPM, the MARB phenotypes and the according behavioral genes are exchanged. In runs including the full ceGPM model, additionally the MAPTs and the corresponding translator genes are exchanged.

Fitness calculation. Fitness calculation is done basically in the same way as in the previous chapter, using the GP fitness calculation in the first and third part of the experimental study described below, and the CA fitness calculation in the second part. The fitness snapshot for GP is calculated slightly differently than before, as given in Alg. 5.2. There, instead of giving a bonus for driving, a penalty is given for not driving with the purpose to increase selection pressure towards passing the gate, as positive fitness cannot be achieved anymore by solely performing CA.

---

input : Current operation $op \in Op$ of a robot R at a time step t; number of collisions $|Coll|$ occurred since last snapshot before t; Boolean value $Gate$ indicating if the gate was passed since the last snapshot before t.

output: Fitness snapshot $snap_{GP}^R(t)$ for R at time step t.

int $snap := -3 \cdot |Coll|$;

if $op \neq (Move, X, .)$ *for all* $X \in B$ **then**
| $snap := snap - 1$;
end

if $Gate$ **then**
| $snap := snap + 10$;
end

return $snap$;

Algorithm 5.2: Computation of a fitness snapshot $snap_{GP}^R$ for GP behavior.

Every robot starts with zero fitness at the beginning of a run, to which the respective fitness snapshot is added every t_{snap} simulation steps. Every t_{evap} simulation steps, evaporation is performed, dividing fitness by $E = 2$. As before, mutation does not change the fitness value, after reproduction, the fitness of the selected parent is adopted, and after application of the memory genome, the stored fitness value is reassigned.

5.3 Evaluation of the Proposed Evolutionary Model

In this section a study is described which uses the examples of GP and CA behavior to evaluate the performance of the proposed model in comparison to other evolutionary approaches described in this thesis. In the first part, runs based on the best setup found for the static approach proposed in Chap. 4 are compared to a variety of runs using different parameter combinations for the new ceGPM-based approach as well as for an intermediate method where the universal translator u (see above) is used as a fixed GPM, but mutation is performed by the new mutation operator M_G on the genotype level. In the second part, originating from the results of the first part, an improved version of the newly proposed ceGPM is introduced that is compared to the first version in terms of evolvability. In the third part, a more detailed comparison of the proposed approaches is performed, including the two ceGPM-based approaches and the according two intermediate approaches, all four combined with a study of the effects of the memory genome.

5.3.1 First Part – Method of Experimentation

In the first set of experiments, a fixed and a completely evolvable version of the new approach introducing mutation on the genotype level are investigated in comparison to each other and to the fixed GPM-less model from Chap. 4. As a test case, evolution of GP behavior is used in all runs. The runs are kept preferably simple with respect to the other evolutionary operators involved to focus on the effects of the ceGPM and the new fixed GPM, respectively. Particularly, no recombination operator and no memory genome, as proposed in Chap. 4, are investigated.

As in Chap. 4, all runs have been performed on a rectangular field sized 1440 mm× 980 mm with a gate sized 190 mm in the middle (cf. Fig. 5.7). At the beginning, 30 robots are distributed uniformly randomly in positions and angles on the field, equipped with empty behavioral automata and the universal translator u depicted in Fig. 5.3b. The runs last for 300,000 simulation steps (which is about 55 minutes in a real-world scenario based on the calculation proposed in Chap. 4). The following three groups of runs have been performed for which Tab. 5.1 shows the detailed parameter setups:

- Setting 1: ceGPM with mutation M_G on the genotype level, as proposed in this chapter. The parameter ϵ^{trans} varies in 50 equal steps from 0.01 ‰ to 2.00 ‰. For each ϵ^{trans} value, 10 runs with different random seeds have been performed.
- Setting 2: Fixed GPM with mutation M_G on the genotype level, modeled as the approach in setting 1 with $\epsilon^{trans} = 0$. This setting was tested with 50 different ϵ values (10 runs each) in a preliminary study, and is set here constantly to the best found value of 5 ‰.
- Setting 3: No (explicit) GPM with mutation operator M. 500 equal runs with the best setup found in Chap. 4 have been performed.

For reproduction, $p = 8$ parents are used in all settings as this number has been found to be among the best in Chap. 4; with respect to statistical significance, $p = 6$ or $p = 7$ would have been equally justified. As before, all experiments have been performed using simulated Jasmine IIIp robots (cf. Figs. 4.6a and 4.6b) in a two-dimensional simulation environment implemented in EAS (cf. Chap. 3). Movement, sensor calculation,

Table 5.1. Parameter setups of the first part of the study. Setting 1 implements the proposed ceGPM, testing 50 different values for ϵ^{trans}; setting 2 is a special case of setting 1 with no mutation on the translator level, i. e., a fixed universal translator is used during the complete runtime (this can be modeled by setting $\epsilon^{trans} = 0$); setting 3 is the best found setup from Chap. 4 without an (explicit) GPM. In the table, the sign "‰" denotes "per mill" meaning $1/1000$.

Parameter	Setting 1	Setting 2	Setting 3
t_{mut}	100	100	100
t_{mut}^{trans}	1000	-	-
ϵ	5 ‰	5 ‰	-
ϵ^{trans}	$0.01, \ldots, 2$ ‰	-	-
d	2	2	-
d^{trans}	1	-	-
$f_+(\lvert g \rvert)$	$20 - \lvert g \rvert/30$	$20 - \lvert g \rvert/30$	-
$f_-(\lvert g \rvert)$	$1 + \lvert g \rvert/30$	$1 + \lvert g \rvert/30$	-
$f_+^{trans}(\lvert g \rvert)$	10	-	-
$f_-^{trans}(\lvert g \rvert)$	10	-	-
p	8	8	8
t_{snap}	50	50	50
t_{evap}	300	300	300
Runs performed	10 per ϵ^{trans} value	10	500

collision simulation and other related technical simulations of the robot have been performed in the same way as in Chap. 4, cf. page 96.

5.3.2 First Part – Results and Discussion

Overall, the experimental results imply that an improvement in terms of fitness and complexity of the behavior can be achieved when using either the completely evolvable or the fixed GPM (settings 1 and 2) compared to setting 3 without an explicit GPM. Furthermore, the ceGPM seems to outperform the fixed GPM in terms of complexity of the evolved behaviors. In the following the results are discussed in detail.

Resulting MARBs. As in the previous chapter, most of the evolved MARBs can be categorized with respect to the following three groups (cf. page 117):

1. *"Collision Avoidance"*: driving around arbitrarily (mostly straight forward) as long as no obstacle is blocking the way and bypassing obstacles otherwise;

2. *"Altruistic Gate Passing with Collision Avoidance"*: the same as "Collision Avoidance" as long as the gate is out of reach, identifying and passing it when it gets in sensor range, and continuing with "Collision Avoidance" afterward;

3. *"Egoistic Gate Passing"*: the same as "Altruistic Gate Passing with Collision Avoidance", except for the case when the gate has been found; in that case a robot oc-

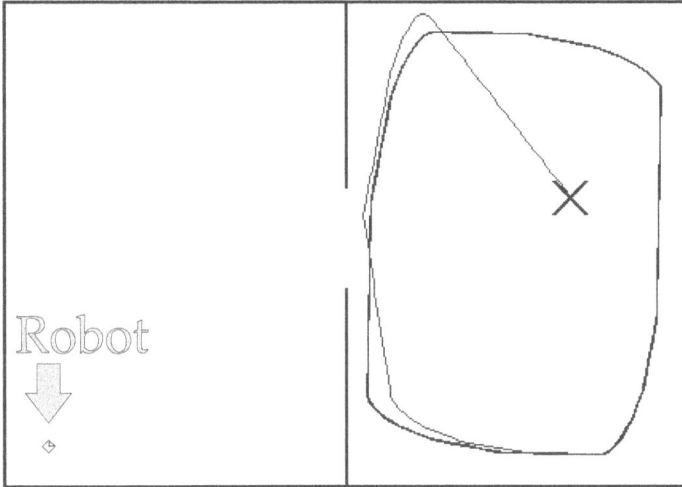

Fig. 5.7. Trajectory of an evolved MARB from group 1 (Collision Avoidance). The robot has an obstacle avoidance capability, but it cannot pass the gate. "X" marks the starting spot, a robot is drawn to scale in the lower left corner.

cupies it in a greedy manner and continues passing it back and forth blocking it for other robots; in some cases two robots share the gate, passing it constantly.

Generally, all observed populations involve MARBs from one of the three groups as CA has been evolved robustly, meaning that all populations lacking a GP behavior at least include MARBs from group 1 (according to the "non-zero fitness" measure introduced in Chap. 4). Fig. 5.7 shows the trajectory of an evolved MARB from group 1, executed on a single robot in an empty field. The robot is able to avoid collisions with walls, but it cannot identify and pass the gate. Behaviors from this group are of minor complexity as CA can be performed by a rather simple two-state MARB in a purely reactive manner. Individual fitness and population fitness are lower than in groups 2 and 3, since the gate passing bonus is hardly ever achieved. Behaviors from this group are mostly undesired as they represent only an intermediate step on the evolutionary path to GP.

Group 2 involves more complex MARBs which allow a gate detection and represent the highest level of adaptation to the desired GP behavior. Here, the whole population is capable of exploiting the gate bonus leading to a high population fitness. An example behavior from group 2 is shown in Fig. 5.8 where a Wall Following behavior has been evolved, meaning that every robot is driving parallel and close to the wall following all its bendings. At the position of the gate, Wall Following automatically leads to a gate passage as the robot follows the pathway of the wall. Therefore, all robots drive in a line and pass the gate twice during a circumnavigation of the field (with the exception of occasional outliers). The version of Wall Following behavior in this figure

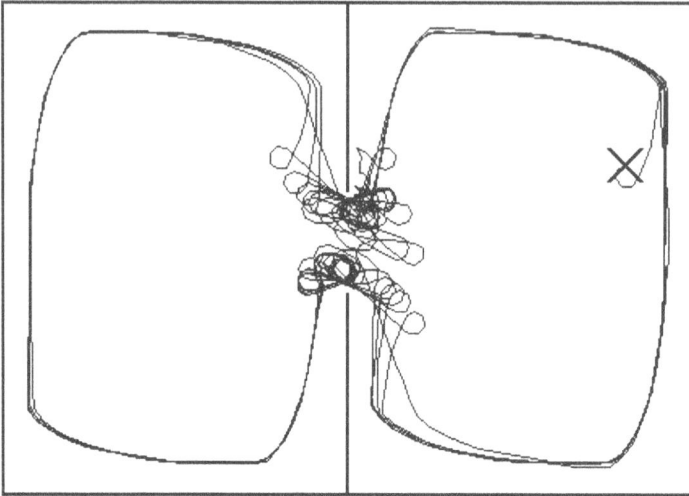

Fig. 5.8. Trajectory of an evolved Wall Following MARB from group 2. The robot follows the walls in the field, thus passing the gate twice every circumnavigation; when confronted with more obstacles such as other robots, the robot is capable of avoiding collisions. Turning right is done by turning left for about 270 degrees in little circles which makes the according automaton surprisingly simple (cf. Fig. 5.9).

is particularly interesting as it is produced by the very simple automaton shown in Fig. 5.9 (tautological parts of conditions have been removed). The automaton does not

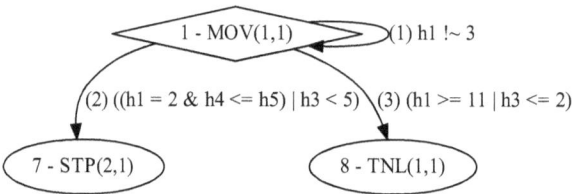

Fig. 5.9. Surprisingly simple evolved Wall Following MARB. Executing this MARB, a robot follows a wall without using any *TurnRight* operations, by only turning left in small circles until a desired driving direction is reached (cf. trajectory in Fig. 5.8).

even have a state with a *TurnRight* operation, but it emulates turning right by turning left by a large amount (cf. the little "$^3/_4$-circles" in the trajectory) until the robot faces in the desired direction.

Behaviors from group 3 have a similar complexity as those from group 2 as they, too, involve detecting and passing the gate. In many cases, this behavior has been

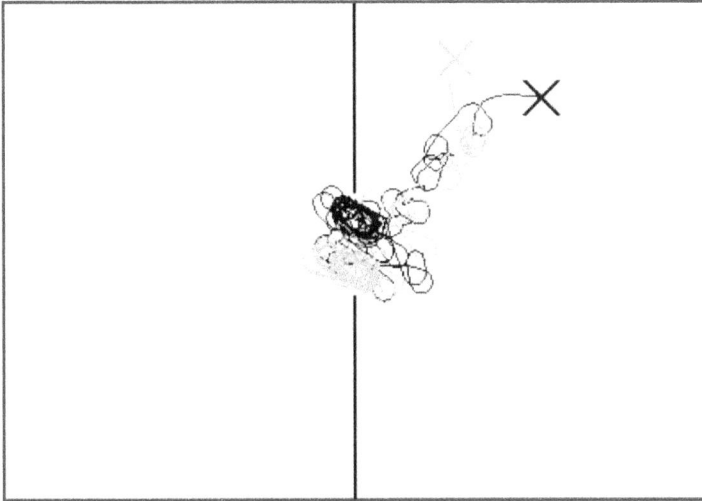

Fig. 5.10. Trajectories of an evolved MARB controlling two robots to pass the gate constantly (group 3). At the beginning, the robots drive more or less randomly in little circles until they find the gate. Then they start to pass the gate back and forth without disrupting each other.

evolved to a robust state where the gate is constantly occupied with one or two robots driving back and forth. However, these behaviors are only capable of exploiting the gate bonus for a small number of robots at once (two is the highest number ever observed) which pass the gate constantly and, therefore, have an extraordinary high fitness. The population fitness, however, is lower than in group 2, since the other robots cannot pass the gate at the same time. In some cases, CA has been evolved as a part of the egoistic gate passing behavior, but mostly the robots drive rather randomly until one or two of them find the gate. Fig. 5.10 shows the trajectories of two robots with MARBs from group 3, quickly finding the gate and passing it constantly henceforth.

As before, due to the high number of runs, not all populations have been observed individually. Therefore, it has been found evident to define populations to belong to groups 2 or 3 if the number of gate passages per time exceeded a certain threshold. By setting the threshold to 0.8 gate passages per 100 simulation steps, this automatic categorizations matches the observations fairly well. Based on this procedure, 45 % of the runs using the new approach with both fixed and flexible GPM, i. e., settings 1 and 2 belong to these groups; on the other hand, only 29 % of the non-GPM runs, i. e., setting 3, belong to these groups. A more detailed comparison between groups 2 and 3 is not provided as they fade into each other and are often hard to distinguish even by observation.

Behaviors from groups 2 and 3, while being significantly more complex than those from group 1, are still far from being "truly complex". However, behaviors such as the

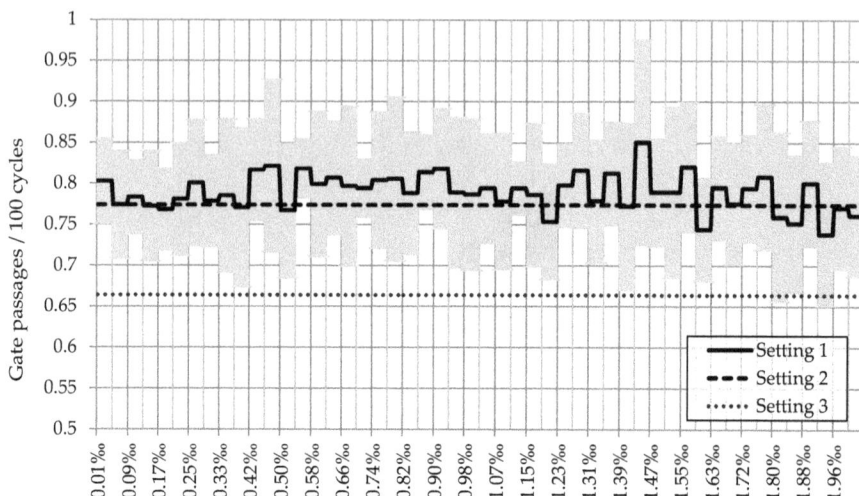

Fig. 5.11. Number of gate passages of the three settings. Average number of gate passages per 100 simulation steps in the three settings, plotted for the 50 different ϵ^{trans} values of setting 1 (X axis). The gray bars denote the SD of the values for setting 1. The SD of setting 2 is similar, the SD of setting 3 significantly smaller; both are omitted in the figure for a clearer representation.

surprisingly simple Wall Following in group 2 or the two-robot gate exploitation in group 3 by itself can justify the utilization of an evolutionary approach rather than an engineering process, as it is hardly imaginable for humans to come up with such solutions.

Overall Results. Fig. 5.11 shows the mean gate passages per 100 simulation steps during the runs of setting 1 (solid line), setting 2 (dashed line), and setting 3 (dotted line), plotted for the 50 different ϵ^{trans} values of setting 1. The X-axis is labeled with the values for ϵ^{trans} in per mill (every other value is omitted for a better readability). In the same way, Fig. 5.12 shows the mean fitness of the runs. Apparently, all runs with an explicit GPM outperform the runs without in terms of the average number of gate passings and mean fitness. This confirms the hypothesis stated above that an explicit GPM can improve the performance of evolution.

Moreover, Fig. 5.11 shows that the number of gate passings is greater with ceGPM than with fixed GPM for most values of ϵ^{trans}, on average 0.79 vs. 0.77 per 100 simulation steps. E. g., for ϵ^{trans} between 0.42 and 1.15, in all but one case the ceGPM outperforms the fixed GPM. This indicates that slightly more complex behaviors evolved with the ceGPM than with fixed GPM. However, due to variance, the difference in performance is too small to be considered significant as is indicated by the rather high standard deviation of the runs from setting 1. At this point it is only a conjecture that

Fitness

18
16
14
12
10
8
6
4
2
0

Setting 1
Setting 2
Setting 3

0.01‰ 0.09‰ 0.17‰ 0.25‰ 0.33‰ 0.42‰ 0.50‰ 0.58‰ 0.66‰ 0.74‰ 0.82‰ 0.90‰ 0.98‰ 1.07‰ 1.15‰ 1.23‰ 1.31‰ 1.39‰ 1.47‰ 1.55‰ 1.63‰ 1.72‰ 1.80‰ 1.88‰ 1.96‰

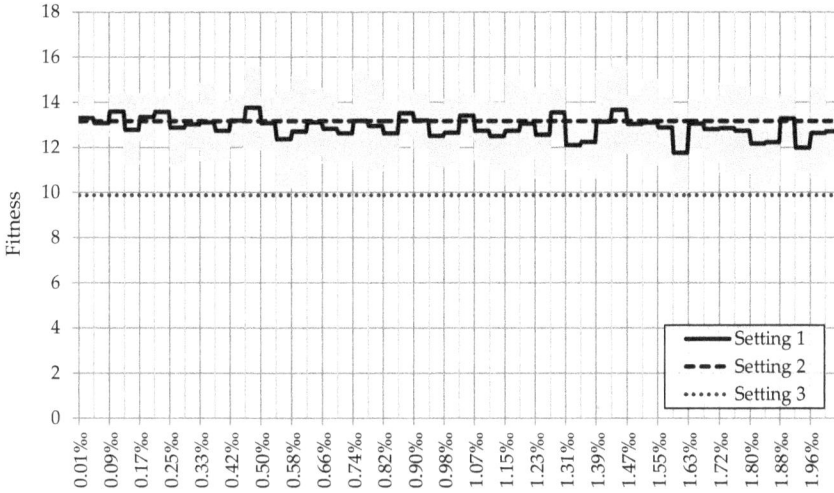

Fig. 5.12. Fitness of the three settings. Average fitness in the three settings, plotted for the 50 different ϵ^{trans} values of setting 1 (X axis). The gray bars denote the SD of the values for setting 1. The SD of setting 2 is similar, the SD of setting 3 significantly smaller; both are omitted in the figure for a clearer representation.

the results confirm the predicted effect of a higher complexity due to the usage of the ceGPM. However, two observations support the conjecture:

1. Over a wide range of ϵ^{trans} values (overall 38 out of 50 values), setting 1 performs best in terms of average gate passings per 100 simulation steps.

2. As Fig. 5.12 shows, mean fitness is mostly lower for setting 1 than for setting 2 (on average 12.86 vs. 13.16). This can be explained by a higher diversity in the populations, due to two concurrent mutation operators, which leads to a higher probability for disappearance of already learnt good behavior. Therefore, in runs with low mean fitness, the counted number of gate passings must have been achieved during shorter periods of good behavior than in runs with high mean fitness. Thus, low mean fitness combined with a high number of gate passings can account for a high complexity of behavior which has been present for a short time.

Nevertheless, these observations are both based on non-significant data and cannot confirm the conjecture. Moreover, as the experiments in Sec. 5.3.3 show, evolvability of the ceGPM used here is rather poor, suggesting that the improvements observed are mainly caused by a higher population diversity due to the two new mutation operators.

Overall, the introduced genotypic representation, both with fixed and evolvable GPM, improves the performance of evolution compared to the former approach without an explicit distinction between genotype and phenotype. There are indications that within the range of tested ϵ^{trans} values an optimum can be found where the ceGPM significantly outperforms the fixed GPM. A more detailed analysis of a subrange of these values is given in part 3 of this study (cf. Sec. 5.3.6) where, however, such an optimum cannot be found, cf. discussion there.

5.3.3 Second Part – An Alternate Completely Evolvable Genotype-Phenotype Mapping and its Effects on Evolvability

As shown in the previous section, the ceGPM based on the MAPT model proposed at the beginning of this chapter is capable of significantly improving evolutionary success in terms of fitness and behavioral complexity compared to an approach without any explicit GPM, and, to a smaller amount, compared to an approach with a fixed GPM. However, it has not been explored whether or not the reason for this success is indeed an increase in evolvability during the run as predicted in the beginning. An alternative explanation could be that the ceGPM produces a higher population diversity improving the search space coverage and, thus, evolutionary success. In this section, an improvement to the MAPT model is suggested and the new version is compared to the old one with respect to their effects on evolvability. For this purpose, an objective measure of evolvability is defined and applied to the MAPT model from the previous section as well as to the newly proposed model MAPT′. In a broad range of parameter setups, performing evolution of CA as a benchmark behavior, the quality of evolvability of the two approaches and its improvements during a run are studied.

The MAPT′ model. The new ceGPM proposed here extends the MAPT model by adding a new command to the script language to solve a structural problem of the MARB and MAPT models. As described above, both models resolve the case that a state has no active outgoing transitions by redirecting the automaton to the initial state, cf. Fig. 4.1 on page 71. This is done to avoid deadlocks in cases where no outgoing transition can ever become active, e. g., if there exists no outgoing transition at all. A consequence of this procedure is a great improbability of evolving an automaton which has interconnected subparts for certain subbehaviors. Since the automaton jumps to the initial state from any state without an active transition, subparts which do not involve the initial state are typically visited only briefly. This, however, leads to behaviors (and, more importantly, in the continued course of evolution to translators) that are close to being purely reactive as they cannot store information about the past by remaining in a certain state or set of states. It is, however, essential to both models to be capable of making decisions in a more than purely reactive way. For MARBs, structures are required which react differently upon environmental influences depending on the past experiences to allow for complex behavior. This capability is largely lost if the automaton restarts the behavior every few simulation steps. For MAPTs, this ability is even more essential as even a simple translation process involves decisions depending on decisions made earlier. For example, the manually constructed universal translator u is fully connected and has essential memorizing capabilities built into its states, such as being in condition construction mode as long as the currently processed genotype symbols do not indicate that the condition is finished. It is hard to imagine translators that lack a memorizing capability to be capable of achieving other than very simple GPMs.

The new script command implemented to overcome the problem of jumping too frequently to the initial state is called CPL ("complete outgoing transitions"), and it is used in combination with two parameters $X, Y \in B_+$ as a new operation

$$CPL(X, Y).$$

It induces the insertion of a transition from X to Y using the same repair process as $EDG(X, Y)$, and it associates a condition to the new transition which is *true* if and only if all other outgoing transitions of X are inactive.

Let the outgoing transitions of X have the conditions $c_1, \ldots, c_n \in C$; then the new condition associated to the transition inserted by CPL is

$$c_{cpl} =_{def} \bigwedge_{i=1}^{n} \neg c_i.$$

Obviously c_{cpl} is *true* if and only if none of the c_i are *true*. The completing transition introduces a possibility to add an outgoing transition to a state which prohibits an implicit jump back to the initial state from that state. It can be used to establish a subpart of an automaton which is sufficiently interconnected to implement a subbehavior. In addition to adding this operation to the script language, the new initial universal translator u' of the MAPT$'$ model already contains a state producing the new operation. Fig. 5.13 shows u' which includes a new state 5 (not existing in u, i.e., deleting state 5 including all its adjacent transitions from u' yields u), labeled with the new script operation. As state 5 is visited only in those cases that were undefined in u, meaning that the new transitions are all completing transitions for their respective source states, u' still implements the encoding proposed in Chap. 4 as long as syntactically correct genotypes are translated. Only for non-syntactical cases a translation by u' can differ from a translation by u leading to the insertion of completing transitions. However, as all incoming transitions of state 5 are labeled $h_{99} < 1$, this state is visited rather rarely by this initial translator. In the course of evolution, translators can evolve that use the new operation more frequently.

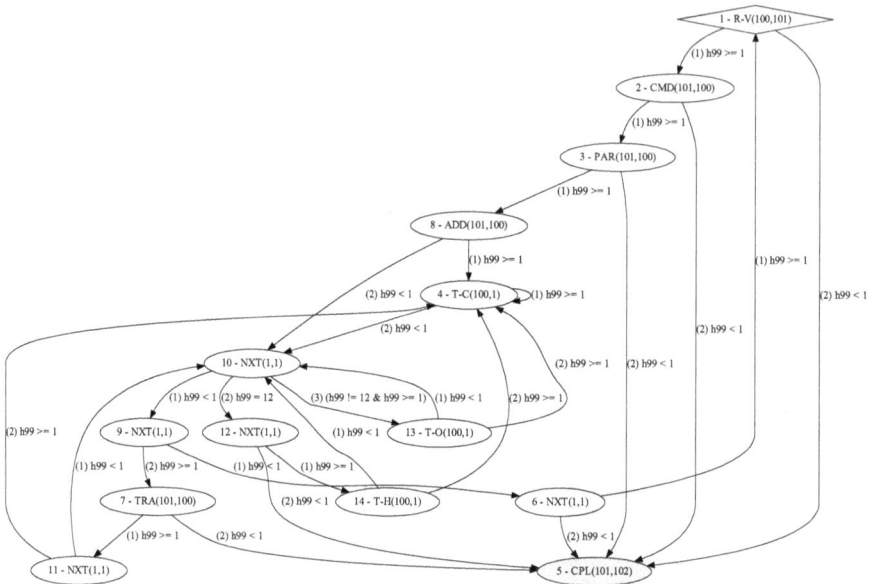

Fig. 5.13. Universal translator u' for the MAPT$'$ model. The only difference between u and u' is state 5 (bottom-right) which uses the new script operation $CPL(X, Y)$. It is visited in all the cases undefined in translator u where the situation that no active transition exists would have redirected the automaton to the initial state – i. e., all the incoming transitions of state 5 are themselves completing transitions of their respective source states. A depiction of translator u' visualized by FMG can be downloaded from http://www.aifb.kit.edu/images/6/6b/Utrans.pdf; there, a variable thickness of the arrows indicates the likelihood for a transition to be taken.

Generating the completing condition. As no $\neg$ operator is part of the language of conditions, negations are not directly representable which leads to the above formula for c_{cpl} not being directly representable. However, it can be created by pushing negation down to the atomic parts of a condition using equivalence relations such as *De Morgan's rule*:

$$\neg(A\ OR\ B) \equiv (\neg A\ AND\ \neg B),\quad \neg(A\ AND\ B) \equiv (\neg A\ OR\ \neg B).$$

At the atomic level, negation can be expressed by relations of the form

$$\neg(a > b) \equiv a \leq b.$$

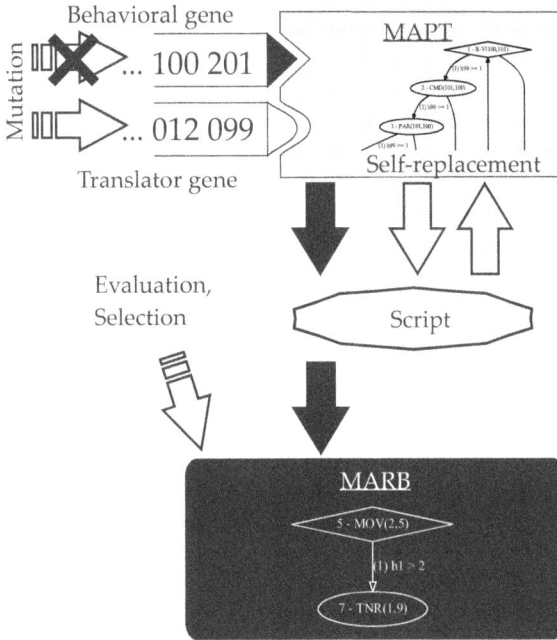

Fig. 5.14. Schematic view of the evolutionary process used for studying evolvability. As before (cf. Fig. 5.4), the mutation operator M_G is active for the translator gene, but it is turned off here for the behavioral gene. All other parts remain unchanged: Mutation triggers the retranslation of both genes replacing the current translator with the retranslation of the mutated translator gene (gray arrows), and the MARB with the retranslation of the unmutated behavioral gene (black arrows).

Using the MAPT$'$ model in an evolutionary run. In principle, an evolutionary run with the new model can be performed exactly as before by just replacing u with u' as the initial translator; this is done in the third part of the study cf. Sec. 5.3.6. However, as opposed to a real-world application where the intention is to mutate both the behavioral and the translator gene, in this section the main focus is on studying evolvability when using ceGPMs. Therefore, a simplified course of evolution as depicted in Fig. 5.14 is used for the experiments. To measure translator evolvability as defined below, behavioral mutations are deactivated to allow for an investigation of pure translator adaptation to the fitness landscape.

The following course of evolution is performed in this section. At the beginning of an evolutionary run, the initial MAPT $u'^0 =_{def} u'$, a self-reflexive translator genotype $g_{u'}^0$ (i.e., it holds that $dec^{trans}(g_{u'}^0, u'^0) = u'^0$), and a randomly generated behavioral gene g_b^0 are given. As a first step, u'^0 decodes g_b^0 into a script which is interpreted to generate a behavioral automaton b^0. This defines the initial behavior which is executed and evaluated in an environment. During the run, random mutations can occur at a simulation step i and turn the current translator genotype g_t^i into some genotype $g_t^{i+1} =_{def} M_G(g_t^i)$. This triggers a process which replaces the current translator t^i by a new translator t^{i+1}, generated by translating the mutated translator genotype g_t^{i+1} with the old translator t^i, i.e., $t^{i+1} =_{def} dec^{trans}(g_t^{i+1}, t^i)$. Afterward, the unchanged behavioral genotype g_b^0 is translated by t^{i+1} to $b^{i+1} =_{def} dec(g_b^0, t^{i+1})$, thus changing a robot's behavior without changing the behavioral genotype. Overall, the procedure is identical to the version used for actual evolutionary applications as described before, except for the omitted mutation of g_b^0 which can, for example, be modeled by a very large mutation interval $t_{mut} \to \infty$.

In this way, translator diversity can be achieved in the population, and, as different translators produce different behaviors from the same genotype, behavioral diversity and adaptation towards a desired behavior can emerge despite of the non-changing behavioral genes. Therefore, evolution is intended to induce an improvement in population fitness by searching on the level of translators instead of directly searching for a good behavior. As behaviors are not mutated directly, the fitness values are a measure of the translators' qualities with respect to an adaptation to a given fitness landscape.

A measure of evolvability. Evolvability is defined to be an adaptation of an evolving system to the search space not only with respect to the behavior evolved, but also to an improvement of behavioral and, in the case of complete evolvability, GPM mutations (cf. Sec. 5.2.1). In the context of ceGPMs an evolutionary system can, thus, be called evolvable if other GPMs (which control the adaptation of mutation) than those given at the beginning of a run are favored in terms of fitness, and if such GPMs are actually evolved during a run. In other words, a "good" ceGPM is expected to be capable of adapting to the behavior being evolved, as otherwise a significantly lower long-term fitness is expected to be achievable. Therefore, an evolvability measure should be based on a setting where behavioral mutations alone strictly cannot generate improvements, but have to be supported by the evolution of the GPM. This setting is created here by turning off behavioral mutations making translator mutations the only alterations that can affect the genotypes. Consequently, a measure of the GPM's participation in the evolution of a behavior is given by the long-term changes in fitness during a run. An increase in fitness can be interpreted as a successful adaptation of the GPM to a given behavior as it can only result from a change of a translator u or u', respectively, to another translator t. If, on the other hand, fitness does not change significantly or if it even decreases during a run, this can be interpreted as no adaptation of the GPM. However, only long-term development of fitness can be considered in this respect as short-term changes can occur without influences of the GPM due to redistributions of existing behaviors or due to uncertainties in fitness calculation.

Therefore, a *long-term increase in fitness* is defined to be an increase of 3 points on a moving average over the average of the fitness values of all robots in a population using a window of 50,000 simulation steps. As the runs last for 275,000 simulation steps, this can be considered a large window, and an increase of 3 points can be considered a high threshold, both assuring that short-term variations in average fitness can be neglected. Fig. 5.15 shows two typical fitness curves of runs which do

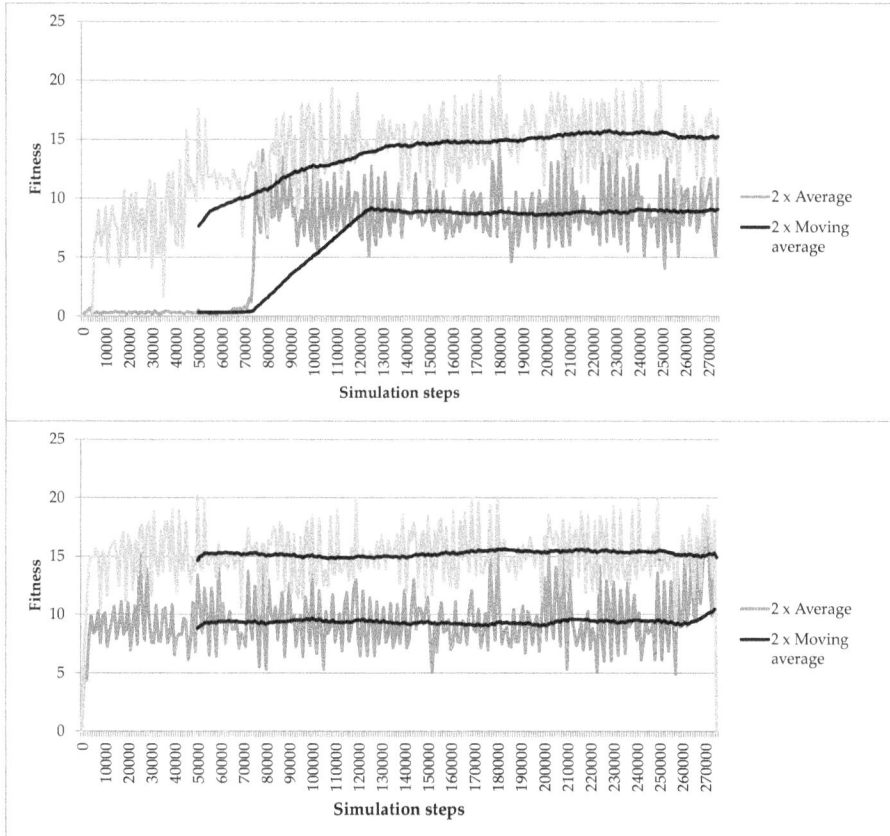

Fig. 5.15. Typical fitness curves indicating (top) and not indicating (bottom) evolvability. The curves in the top chart indicate evolvability due to the increasing moving average, shown in black, by at least 3 points during a run. The curves in the bottom chart indicate no evolvability as the moving average stays at a more or less constant level. The absolute height of the curves is not taken into account as behaviors of different qualities can exist from the beginning in a population.

(top chart) and do not (bottom chart) indicate evolvability, respectively. The quick increase in average fitness at the beginning of three of the four curves originates from redistributions of existing behaviors and the fact that all robots start with zero fitness. Both these effects have hardly any influence on the moving averages shown in red.

Further justification. Reisinger et al. argue that evolvability measures should not rely on benchmark tests only, since evolvability is a general property which should be present in a set of behaviors [164]. In spite of that, the measure proposed here is based on a benchmark test using CA behavior. However, improvements of the behavioral part of the controllers are not directly regarded an indication of evolvability, but only indirectly by their implications concerning improvements of the translators, which represent the underlying mechanism to achieve evolvability. Therefore, the measure can indicate that a translator is capable of adapting domain-specifically during an evolutionary run which, in turn, makes the simultaneous evolution of behavioral automata adaptable. Furthermore, despite evolving a simple, purely reactive CA behavior on the surface, the alongside evolved translators can presumably be neither simple nor purely reactive if they are assumed to be capable of performing a complex translation process. Therefore, by the example of evolving CA, general implications can be derived about the quality of evolvability in an evolutionary system. Particularly, the proposed measure of evolvability matches the definition of evolvability given here as well as the common requirements given in the literature, as the remarks by Reisinger et al. do not seem to be applicable, at least to a full extent. Moreover, the results show that the measure is sufficiently precise to differentiate the qualities in evolvability between the MAPT and MAPT$'$ models (cf. Sec. 5.3.5). Nevertheless, studies with different benchmark behaviors, especially more complex ones, may lead to a more specific statement about evolvability.

5.3.4 Second Part – Method of Experimentation

An initial set of experiments – the *first group* – has been set up to cover a broad range of 324 different parameter combinations. Each of these combinations has been tested twice, using different random seeds, for (a) the MAPT model and (b) the MAPT$'$ model, i. e., $2 \cdot 2 \cdot 324 = 1{,}296$ runs have been conducted overall. Those setups from the first group that indicated evolvability according to the proposed measure in at least one of the two identical runs have been called *Possibly Successful Setup (PSS)* and studied further in a second set of experiments – the *second group*. In the second group, both ceGPMs were tested with all PSSs from the first group, both (a) and (b). In this way, the ceGPMs have been given a chance to perform well with any of the initially found PSSs. Accordingly, the second group has been divided into four categories:
1. MAPT model with its own PSSs from group 1(a),
2. MAPT model with the foreign PSSs from group 1(b),
3. MAPT$'$ model with its own PSSs from group 1(b), and
4. MAPT$'$ model with the foreign PSSs from group 1(a).

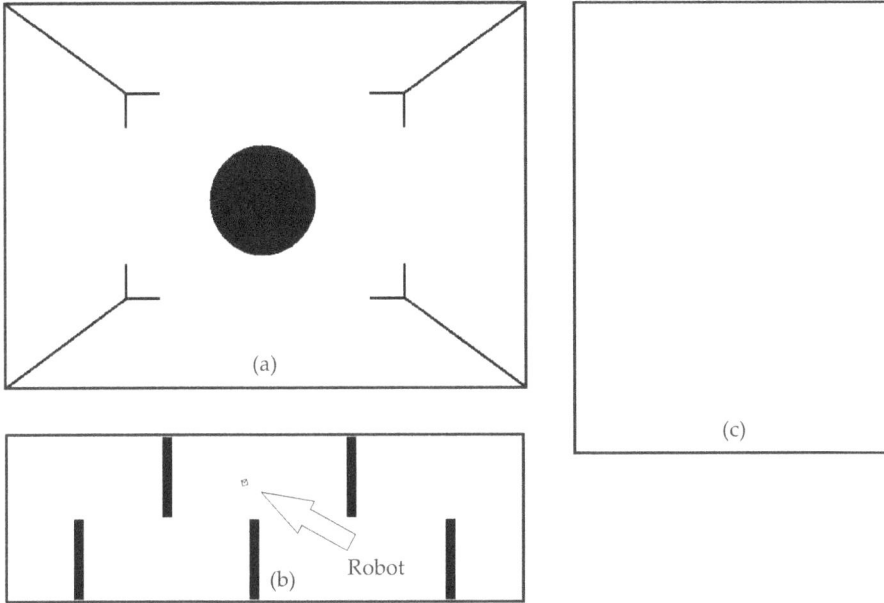

Fig. 5.16. Fields used in the second part of the study. The fields are drawn to scale with respect to each other, the arrow points to a robot drawn to scale accordingly. The fields are called (a) *4-walls*, (b) *5-walls*, and (c) *empty*.

Every run in the second group has been performed ten times with different random seeds to allow for a more detailed analysis. In both groups, the performance of the ceGPMs has been evaluated using the proposed evolvability measure.

Overall setup. Where not stated differently, the runs have been performed in the same way as described in Sec. 5.3.1. In addition to the two different ceGPMs used, five parameters have been altered throughout the setups: the field env, the mutation interval for translator mutations t_{mut}^{trans}, the gene length factor for translator genes e^{trans}, the standard deviation for translator gene mutations d^{trans}, and the number of parents p. In every setup, a swarm of 40 robots is simulated for 275,000 simulation steps in one of the fields depicted in Fig. 5.16. The fields differ in their amount of free space and distribution of obstacles to induce different degrees of environmental selection pressure. The robots are initialized with the universal translator u for MAPT and the universal translator u' for MAPT$'$, the respective self-reflexive gene g_u or $g_{u'}$, and a randomly generated behavioral gene of length 150. Each byte of the behavioral gene is set to the absolute rounded value

$$\min(\text{round}(\text{abs}(x)), 255)$$

of a number

$$x =_{def} rand_{gaussian} \cdot 10$$

i. e., a number drawn randomly by Gaussian distribution with a standard deviation of 10 and a mean of 0, cut off at 255. Mutation is performed repeatedly every t_{mut}^{trans} simulation steps on each individual robot, changing only the translator gene and, as a consequence, the translator which, in turn, changes the behavioral automaton (but not the behavioral gene). Selection is triggered every $t_{rec} = 150$ simulation steps uniformly for all robots. Every robot compares its fitness to the fitness of the $p - 1$ robots closest to itself, adopting both the genotypes and the automata (as well as the fitness) of the robot with the highest fitness. Fitness snapshots are calculated every $t_{snap} = 25$ simulation steps, adding 2 to the fitness of robots being in a *Move*-state and subtracting $3 \cdot |Coll|$ for the number of collisions $|Coll|$ that happened since the last fitness snapshot (note the slight difference to Alg. 4.2 where a reward of 1 has been added for being in a *Move* state; the purpose of the higher reward is to more quickly overcome the state of not moving at all; this alteration is performed by intuition and is not further justified). For evaporation, the fitness values are divided by $E = 1.5$ every $t_{evap} = 95$ simulation steps.

Parameter setups of the first group of experiments. For the five variable parameters, the following values have been tested in a fully factorial way:

- $p \in \{2, 4, 8\}$,
- $env \in \{4\text{-walls}, 5\text{-walls}, \text{empty}\}$ (cf. Fig. 5.16),
- $t_{mut}^{trans} \in \{50, 200, 1000\}$,
- $d^{trans} \in \{3, 9, 27\}$,
- $\epsilon^{trans} \in \{10^{-4}, 10^{-3}, 10^{-2}, 10^{-1}\}$.

Parameter setups of the second group of experiments. The first set of experiments has yielded 12 PSSs for runs with the MAPT model and 45 PSSs for runs with the MAPT′ model indicating a higher evolvability of the new model. Fig. 5.17 shows the 45 PSSs of the MAPT′ model and their performance in the second set of experiments. The PSSs for the MAPT model are not depicted as their performance has been very poor in the second group suggesting exclusively statistical flukes in the first group, see below.

5.3.5 Second Part – Results and Discussion

The performance of the MAPT model in the second set of experiments has been very low in both its own and the foreign PSSs. Of the 120 runs conducted overall in category (1) of group 2 with own PSSs for MAPT, only 4 (3.3 %) have been successful in terms of evolvability; similarly, of the 450 runs in category (2) with foreign PSSs for MAPT, 8 (1.8 %) have been successful. Likewise, the runs with foreign PSSs for the MAPT′ model, category (4), have performed rather poorly being successful in 5 of the 120 runs (4.2 %). None of these setups have yielded more than two successful runs out

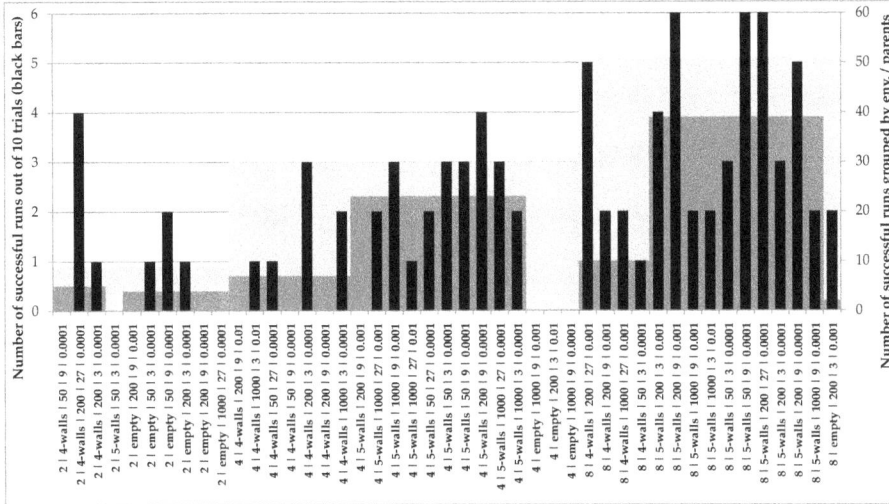

Fig. 5.17. Results of category 3 of the second set of experiments. The chart depicts the category involving the MAPT′ model and its own PSSs. The X axis denotes the PSSs by the setting of their variable parameters in the order: $p \mid env \mid t_{mut}^{trans} \mid d^{trans} \mid e^{trans}$. The runs are sorted by p and then by env. The black bars denote the number of successful runs according to the left Y axis. The gray bars denote the sum of all successful runs grouped by p (light gray) or env (dark gray), respectively, according to the right Y axis.

of ten for a specific PSS, therefore, the results are not analyzed here in detail. In contrast, category (3), containing own PSSs of the MAPT′ model, has yielded 90 (20 %) successful runs. At maximum, six out of ten runs indicate evolvability for a specific PSS in this category, cf. Fig. 5.17. This has been interpreted as an indicator that the MAPT′ model can provide evolvability in certain parameter combinations while this capability has not been found for the MAPT model. However, as the MAPT′ model has not been able to establish evolvability in runs with foreign PSSs, and only for a subset of the own PSSs, the choice of the evolutionary setup seems to be crucial to the success of the ceGPM.

Properties of successful setups. As Fig. 5.17 shows, none of the runs with $e^{trans} = 10^{-1}$ were PSSs for MAPT′. Apparently, this setting involved too large mutations to allow a directed evolutionary search. Of the other parameters, the number of parents and the environment seem to have the greatest influence on evolvability having the best effect at $p = 8$ and $env = 5$-walls. Both these settings induce a high selection pressure as $7/8$ of lower-fitness automata are ruled out at each selection, and 5-walls is the smallest environment with the highest rate of obstacles, providing the greatest punishment for collisions. Overall, this indicates that large mutations tend to let

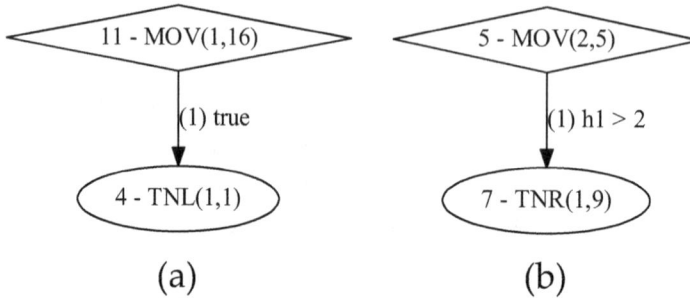

Fig. 5.18. Typical MARBs evolved in a setting without behavioral mutations. (a) Circle driving behavior, mediocre fitness; (b) basic CA behavior, fairly high fitness. According to the proposed evolvability measure, both MARBs can indicate evolvability if they emerge from formerly worse populations.

translators drift away from desired search space regions, which has to be delimited by smaller mutations and a high selection pressure.

Resulting MARBs. Fig. 5.18 shows two structurally typical resulting MARBs (simplified to a minimal representation by deleting unreachable or inactive parts). The left implements a circle driving behavior gaining mediocre fitness as the according MARB is fairly often in a *Move* state and collides rather rarely, since after few initial collisions every robot tends to find a spot where it can drive circles without crossing the trajectories of others. However, the behavior is punished in every other step when it is in a *Turn* state rather than a *Move* state. Variations of this behavior have two *Move* states which leads to driving larger circles and a slightly higher fitness. The right MARB gains higher fitness by performing a true CA behavior. It drives forward as long as sensor 1 has a value less than or equal to 2, which indicates that no obstacles are blocking the way, and turns right otherwise. However, such CA behaviors have been less frequently evolved than the simpler circle-driving behaviors. While this has been expected due to the limited evolutionary operators in this setting, it indicates that the ceGPM has difficulties in evolving more complex MARB conditions. As conditions build a major foundation of the semantics of a MARB, future investigations might focus on improving the generation of conditions by the ceGPM, at least if it becomes apparent that this capability is limited in settings where behavioral mutations are activated.

Both behavioral automata in Fig. 5.18 have probably not been constructed under the use of the new script command *cpl* of the MAPT′ model (although the *true* transition of MARB (a) might have originated from completing state 11 when it had no outgoing transitions). However, in the process of translator evolution the command *cpl* obviously supported the generation of new translators in a significantly better way than the MAPT model, as is shown by the improvement in evolvability. It is expected that the command has frequently been used for the construction of both translator and behavioral automata, although the former usage on its own would suffice to explain the improvement in evolvability. The evolved translator automata are not analyzed here, as their structure is less clear than the structure of most MARBs and as there are so far no general algorithms to automatically analyze MAPTs. Future efforts should involve finding such methods providing a more formal analysis of both MARBs and MAPTs from a semantics point of view. Their existence is a general claim of this thesis given by the theoretical body of FSMs. In the MARB case, Chap. 4 shows some basic ideas how such an analysis can be performed, but more sophisticated methods are desired for more complex automata which particularly includes most MAPTs.

Overall, according to the proposed evolvability measure it has been shown that the presented MAPT$'$ model is capable of performing an adaptation to a given search space while this property could not be demonstrated for the MAPT model. This suggests that the improvements observed in Sec. 5.3.1 may have had other causes than an improvement in evolvability, such as an improved search space coverage. It can be concluded that the MAPT model does not gain evolvability to a statistically significant extent while the MAPT$'$ model does. However, as stated before, a good choice of the evolutionary setup is crucial for the success of an evolutionary system which also applies to approaches using a ceGPM. Here, however, this means that a good evolutionary setup has to be found only once for a set of various behaviors, as the purpose of a ceGPM is to adapt to different behaviors being evolved. The results indicate that the evolutionary setup in a ceGPM scenario must have specific properties to actually allow for an improvement in evolvability. Deviations from these settings seem to critically impair the evolutionary results which, on the other hand, suggests that even better performances can be expected with more suitable evolutionary setups. In the third part of the experiments section, a more detailed analysis of the performance of different evolutionary setups is conducted.

5.3.6 Third Part – Method of Experimentation

The third part of the study aims at gaining more practical information about how the initial setup influences the outcomes of evolution. Using the results from the first two parts, the most promising parameter ranges for the mutation probability of translator genotypes have been tested for the MAPT model as well as the MAPT$'$ model in a rather broad analysis. The best parameters found in this first set of experiments have been studied in a second set of experiments in more detail. There, the MAPT model as well as the MAPT$'$ model and the fixed-translator model, using the non-evolving universal translators u and u', respectively, have been tested, and combined with an analysis of the influence of the memory genome. As a benchmark, GP behavior has been evolved in a scenario identical to the experiments in Sec. 5.3.1, where not stated otherwise.

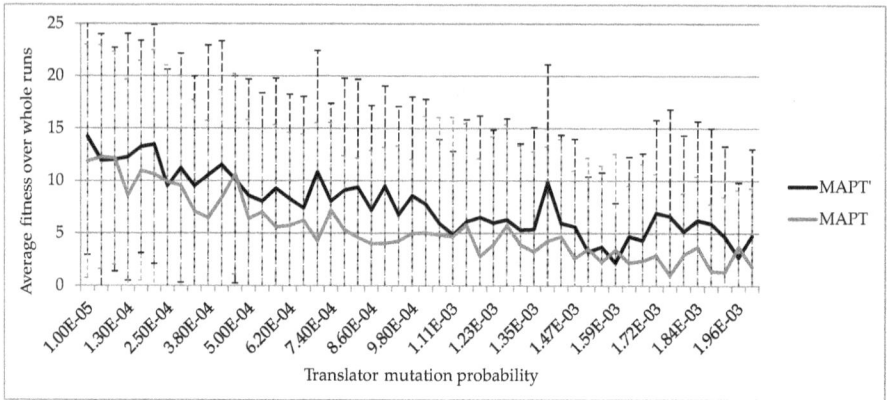

Fig. 5.19. Average fitness over the whole runtime of the first set of runs. For both the runs with the MAPT and the MAPT′ model the chart shows the average fitness values during the whole runtime distributed along the tested mutation probabilities ϵ^{trans} (the caption of the X axis displays only a subset of the overall 50 values). SDs are depicted as error bars, however, they are very high and not particularly meaningful. As all fitness values during runtime are covered, their progression is expected to be volatile, thus, yielding high SD.

The first set of experiments. In the first set of experiments, the MAPT and MAPT′ models have been evaluated with 50 values for the mutation probability of translator genes ϵ^{trans} ranging from 10^{-5} to $2 \cdot 10^{-3}$ in steps of equal size (i. e., $3.98 \cdot 10^{-5}$). The other values have been set to the best values found in the first part of the study. Every parameter combination has been tested in ten identical runs differing only in the underlying random seeds.

The second set of experiments. As the low mutation rates show a better performance in the first set of experiments (cf. Sec. 5.3.7), in the second set the mutation rates with

$$\epsilon^{trans} \in \{0, 10^{-5}, 5 \cdot 10^{-5}, 9 \cdot 10^{-5}, 1.3 \cdot 10^{-4}, 1.7 \cdot 10^{-4}, 2.1 \cdot 10^{-4},$$
$$2.5 \cdot 10^{-4}, 2.9 \cdot 10^{-4}, 3.3 \cdot 10^{-4}, 3.8 \cdot 10^{-4}\}$$

have been tested in more detail, conducting every run 100 times with identical parameter settings and different random seeds. Furthermore, every ϵ^{trans} value has been tested with and without the usage of the memory genome. Note that the $\epsilon^{trans} = 0$ value corresponds to the static translator version introduced in the first part of the study; it is referred to as *static* in the following.

5.3.7 Third Part – Results and Discussion

The first set of experiments. Fig. 5.19 and Fig. 5.20 show the average fitness values

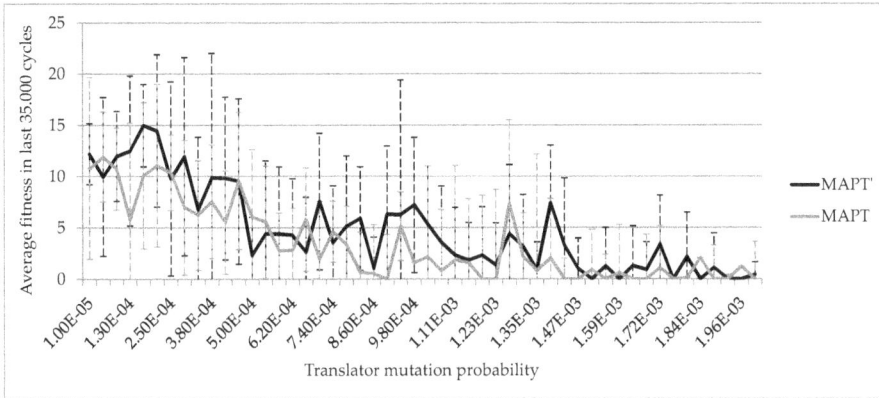

Fig. 5.20. Average fitness in the last 35,000 steps of the first set of runs. For both the runs with the MAPT and the MAPT′ model the chart shows the average fitness values in the last 35,000 simulation steps distributed along the tested mutation probabilities ϵ^{trans} (the caption of the X axis displays only a subset of the overall 50 values). SD is depicted as error bars; it is lower than in Fig. 5.19, showing that fitness tends to converge in the last steps, but it is still rather high indicating that the data is not statistically significant and can be used as a first impression only.

during the complete runs and in the last 35,000 simulation steps of the runs, respectively, divided in runs with MAPT and MAPT′ and plotted for all tested mutation rates. Obviously, for high mutation rates the quality of the behavior in the last population decreases which is not surprising as evolution degenerates toward a random search if there is too much translator mutation involved. Therefore, the chart suggests that further experiments should focus around the lower of the tested mutation rates. Overall the MAPT′ model performs better than the MAPT model (on average 7.73 fitness points vs. 5.43 fitness points over the complete runs, and 4.84 fitness points vs. 3.44 fitness points within the last 35,000 simulation steps); there is a tendency for a peek of the MAPT′ model between $\epsilon^{trans} = 1.3 \cdot 10^{-4}$ and $\epsilon^{trans} = 2.5 \cdot 10^{-4}$. However, SD is high for both curves, and there is a necessity of investigating particularly the disputable low ϵ^{trans} values in more detail. Another observation is that, on average, fitness is significantly higher over the complete runs than in the last population. This can only be explained by short-term increases in fitness due to (expectably) good behavior which disappears before the end of the runs. The second set of experiments investigates if the memory genome is capable to diminish this undesired effect.

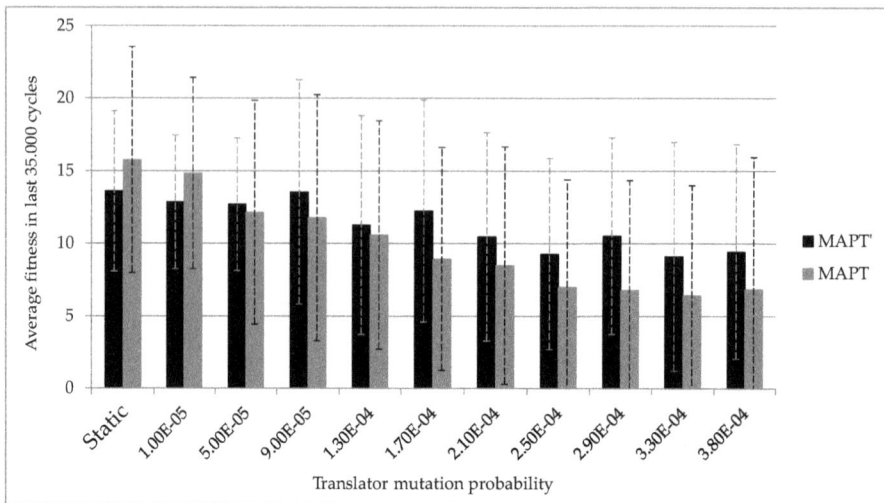

Fig. 5.21. Average fitness in the last 35,000 simulation steps of the second set of runs. The chart corresponds to runs without usage of the memory genome. The error bars denote SD.

The second set of experiments. Fig. 5.21 shows the results of the runs without the memory genome in terms of average fitness in the last 35,000 simulation steps. The chart corresponds to the results shown in Fig. 5.20, and the values basically match the values shown there; the standard deviation is still rather high prohibiting final statements about the quality. However, the peek suggested in the discussion of the results of the first set cannot be confirmed. Rather, the static translator appears to outperform the evolvable translator versions, and further increasing the mutation rate leads to a decrease of fitness in the end of the runs. Surprisingly, using the static version or the very week mutation $\epsilon^{trans} = 10^{-5}$, the MAPT' model is not capable of outperforming the MAPT model while it does show better results for any greater ϵ^{trans} values. Both these observations seem to show that the ceGPM is not capable of improving evolution and that amplifying mutation of the translator (which is also a consequence of the MAPT' model's completing transitions) leads to weaker results. However, as argued in the discussion concerning the first part of this study, fitness and behavioral complexity can be reversely correlated to each other meaning that complex behavior may lead to a lower average fitness. Therefore, the high fitness of the static version does not necessarily mean that this version leads to the most complex and highly adapted behaviors. Furthermore, the second part of this study clearly shows that the ceGPM is capable of improving evolvability which, in turn, can increase the performance of evolution if applied accurately. This indicates that there have to be better evolutionary setups than the ones tested here which lead to an improvement of evolution.

Fig. 5.22 shows the results of the runs which included the usage of the memory

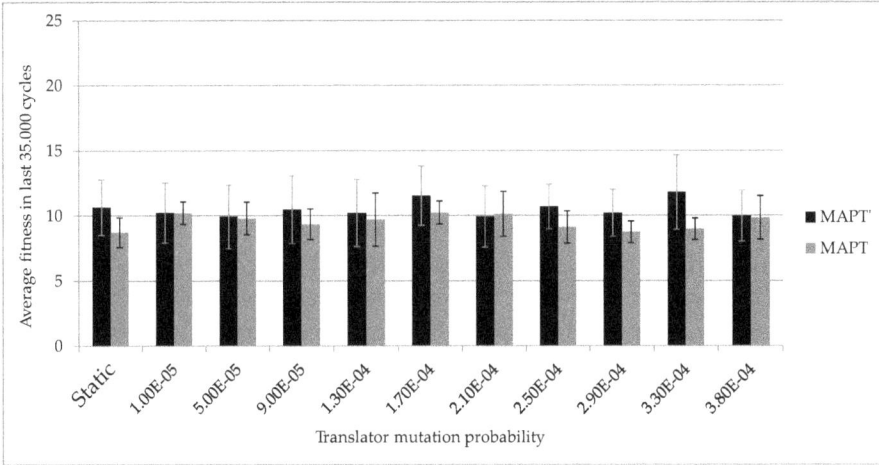

Fig. 5.22. Average fitness in the last 35,000 simulation steps of the second set of runs. The chart corresponds to runs which included the memory genome. The error bars denote SD.

genome. The standard deviation is rather small for these results making them more reliable than any of the preceding results. However, the results are significantly worse in terms of fitness in the last 35,000 simulation steps than in the runs without the memory genome. The memory genome, while preventing a fall-back in bad behaviors after good ones have been found, on the other hand reduces diversity in the population. The problem is apparently that exploitation is pushed to far while exploration does not take place to a sufficient amount anymore. To make better use of the memory genome, more experiments have to be performed including a study of the time interval and the fitness threshold used for that operator. However, as other parameters, too, have an influence on population diversity, the memory genome as a rather lucid operator should be studied after the other parameters have been optimized. Overall, the MAPT′ model is capable of outperforming the MAPT model in practically all parameter setups. This can be explained by the expected beneficial influence of the completing transitions introduced by the script command *cpl*.

5.4 Chapter Résumé

Overall, it can be concluded that much more experiments are required to finally resolve the question of how practically useful the ceGPM can be in general, and, more specifically, how beneficial the MAPT or MAPT′ models can be for ER. As the standard deviation remains high even for the experiments including 100 runs per setup, and as there are many more parameters that have not even been investigated within the large set of runs performed for this chapter, it has to be stated that this final resolution is beyond the scope of this dissertation thesis. Particularly, evolution of behaviors more complex than the ones known from literature has not been accomplished so far.

Nevertheless, there are strong indications that the ceGPM can provide improvements to an evolutionary run if it is utilized accurately. Firstly, the results imply that more complex behaviors have been evolved using the ceGPM than without it meaning that the ceGPM might prove beneficial in scenarios involving more complex behavioral requirements than those for CA and GP. Moreover, another measure than plain fitness might reveal new insights in the objective qualities of the evolved behaviors. Secondly, the MAPT$'$ model has shown to improve evolvability in a scenario arranged specifically to serve as a test case in part two of the study presented above. Transferring the test case more accurately into a real scenario might lead to the expected improvement in evolvability and, hence, in the overall quality of evolution. Thirdly, the theoretical argument stated in the beginning of this chapter concerning benefits of ceGPMs is valid for ceGPMs providing sufficient flexibility and smoothness. Therefore, if no scenario at all can be found supporting this hypothesis by using the approach proposed here, this should be attributed to the possibly too inflexible FSM-based translator model. As suggested before, a ceGPM based, for example, on ANNs might have benefits over an FSM-based version. On the other hand, as argued in Chap. 4, such a solution would lack the simplicity and analyzability of FSMs. Fourthly, the EE approach followed in this thesis might be unfavorable for a proper functioning of a ceGPM as it implies constraints to selection and fitness calculation which affect the distribution and evaluation of behavioral and translator automata. A centralized and offline evolutionary approach might provide means to more accurately select for desired behavioral and translator properties. Furthermore, the ceGPM is in principle not limited to ER and suitable for many evolutionary approaches. In any case, further effort is planned to be put in finding a ceGPM capable to practically improve evolutionary scenarios within or outside of ER.

6 Data Driven Success Prediction of Evolution in Complex Environments

> Preamble: *In contrast to classic Evolutionary Algorithms which use completely goal-based selection operators, selection in nature is largely environment-based in the sense that there is no global goal populations are pushed toward by an external force. In fact, fitness is defined by biologists by means of the amount of an individual's offspring capable of surviving to adulthood which is just complementary to a classic computer scientist's point of view. With Evolutionary Robotics, a computer scientist has to face reality as it introduces an intermediate case between the two extremes: there is a desired goal toward which selection is supposed to push, but there is also a complex environment that interacts strongly with both selection and fitness calculation. As a consequence, providing an appropriate setup for evolution turns out to be much more challenging than just choosing a good selection operator.*

Providing accurate selection is one of the key problems in finding an appropriate evolutionary setup [142] in ER. Selection is considered accurate if "better" controllers in terms of the desired behavioral qualities have a higher chance of being selected than "worse" ones. Although this is an inevitable precondition to successful evolution, it is in many evolutionary scenarios surprisingly difficult to accomplish. This difficulty can be basically ascribed to the inherent interactions with a complex environment. As it is usually not possible to grade arbitrary evolved controllers detached from the environment the desired task is to be achieved in, it is unavoidable to include the environment in the selection process.

Using an environment to establish the quality of controllers makes ESR more closely related to natural evolution than most classic EC approaches. As illustrated in Sec. 2.1.1, the selection process can consequently be seen from a classic EC or a biological perspective. The theoretical model proposed in this chapter has the purpose of improving the understanding of this combined view on ESR and helping researchers and developers in an early phase of a project to grade the quality of an evolutionary scenario in terms of its likeliness to produce successful individuals. It is based on implicit, i. e., environment-given selection properties and the *Selection Confidence (SC)* of a system, which is a measure of the probability of selecting the "best" out of two or more different behaviors.

From a practical point of view, the model can, for example, be utilized to calculate from simulation or laboratory data the expected success of an evolutionary approach when performed in a real-world environment, cf. Fig. 6.1. As a foundation of the model, a data abstraction method is proposed with the purpose of overcoming the reality gap. Furthermore, the model can be used to theoretically and experimen-

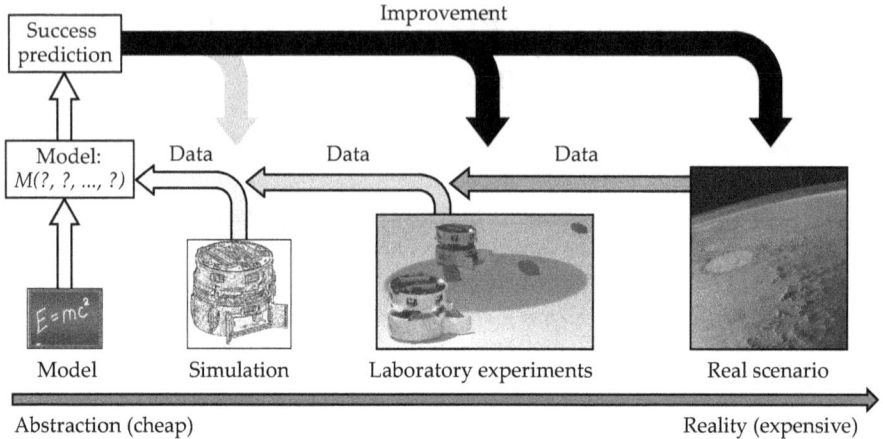

Fig. 6.1. Overview of the intended usage of the success prediction model. The prediction model is fed with data from simulation or laboratory experiments to calculate the expected success of a real-world scenario. In an advanced step, data from real-world experiments can be used to further improve future real-world experiments.

tally study the influence of complex environments on evolution. Particularly, it allows to gradually increase SC from "no confidence", i. e., purely environment-based selection, to "absolute confidence", i. e., always choosing the "best" out of several robots. The model is applicable to evolution in a swarm of mobile robots, or, more generally, to all scenarios where agent behaviors are evolved within a complex environment.

The model presented in this chapter including the experimental results has first been published in 2011 [108]. This chapter extends the paper in several respects, mostly by more comprehensive and profound descriptions and more thorough discussions of the results. Particularly, a novel section about evolution within a purely implicit selection environment (Sec. 6.4.3) introduces a new view on the model's potential impact.

In the next section, the general underlying idea for the prediction model is presented as well as according preliminaries. In Sec. 6.2, a model for completely implicit selection is introduced. Sec. 6.3 presents an extension to the completely implicit selection model to allow for explicit selection to be modeled. Sec. 6.4 describes and discusses several simulation studies to evaluate the introduced models. Sec. 6.5 concludes the chapter and gives an outlook to remaining issues.

6.1 Preliminaries

In this section, preliminary definitions and the general evolutionary algorithm building the foundation of the prediction model are given.

Implicit and explicit selection. The terms *implicit* and *explicit selection* are defined in Chap. 2. Explicit selection is based on an explicit fitness value calculated onboard of a robot or by a global observer. In the following, the term *fitness* is used exclusively for explicit fitness. Implicit selection is induced by the interaction of robots with a complex environment; for example, when reproduction includes spatial proximity, the system implicitly selects for the ability of finding other individuals to reproduce with. In this respect, robots have implicit fitness as well as explicit fitness according to their behavior during a run.

The prediction model is based on a generalized version of the tournament selection-like reproduction procedure proposed in Chap. 4. Rather than requiring spatial proximity between robots for building a tournament, here an arbitrary environmental property is assumed that implicitly selects k robots to join a tournament. Within a tournament, one of the robots is explicitly selected to overwrite the controllers of the $k - 1$ other robots by its own controller. Both selection confidence and the implicit selection property of the environment are parameters to the model. No recombination is considered so far during reproduction for simplicity reasons, however, the model is powerful enough to cover various complex selection schemes.

The evolutionary algorithm. Let E be an environment including a population of n robots. The terms "environment" and "robot" (or "agent") are left loosely defined here, discussions of the rather complex arguments concerning exact definitions in multi-agent systems are given in the literature [203], and some high-level thoughts in Sec. 3.3.6. For the prediction model, the only requirement for an object to be called "environment" is that it has to be a system that at every point in time is in a single state out of a specific, possibly infinite state space. In the following, environments with a spatial extent are considered, and time is assumed to be discrete. Robots are objects that are part of an environment and can get information about the current environment state via a defined set of sensors; furthermore, they can influence the succeeding state by a defined set of actions performed by a set of actuators. Being part of the environment themselves, robots can get information about their own internal state through sensors and change it through actuators.

The process of deciding from sensory data which actions to perform is defined by the *controller* of a robot. Alg. 6.1 shows the evolutionary process which builds the foundation of the prediction model. The algorithm is stated from a population point of view, but it can also be applied in a decentralized way as in [200] (cf. Sec. 6.4). The function *Initialize* places n robots at random positions in the environment E and assigns them some arbitrary (empty, pre-defined or random) controllers from an array of controllers $\vec{\Gamma} = (y_1, \ldots y_n)$. Hereafter, the population of robots can be identified with the population of controllers $\vec{\Gamma}$. $Execute\left(\vec{\Gamma}\right)$ runs the controllers "to perform an atomic action", whatever this means in terms of the controller model (e. g., ANN or MARB). The execution can occur sequentially or concurrently as required for decentralized scenarios. The fitness values of the controllers are stored in $\vec{F} = (f_1, \ldots, f_n)$ and computed by a fitness function f. The function $Mutate$ indicates the mutation operator which is performed repeatedly at time intervals t_{mut}. The function $Match$ selects a set of controllers T of size $|T| = k$ for a tournament. This selection can depend on environmental properties such as spatial proximity. Among the selected controllers T, the controller $index$ is fitness-proportionally chosen by the function $Select$ to overwrite the other controllers in T. Mutation is not explicitly considered in the prediction model, meaning that no specific characteristics of mutations (other than "good" or "bad") are treated. Therefore, the model is assumed to be applied in between mutation operations, predicting the probabilities of beneficial or harmful mutations to spread through the population.

---

input : Population of n initial controllers $\vec{\Gamma} = (y_1, \ldots y_n)$; environment E;
tournament size k; maximum runtime t_{max}; mutation interval t_{mut};
explicit fitness function f.
output: Evolved population.
$Initialize(\vec{\Gamma}, E)$
for $int\ t = 1\ to\ t_{max}$ **do**
    $Execute(\vec{\Gamma})$ // Run controllers
    $\vec{F} := f(\vec{\Gamma})$ // Compute explicit fitness
    if t mod $mutInterval == 0$ **then**
       | $Mutate(\vec{\Gamma})$ // Mutation
    end
    // Mating and tournament selection
    $T := Match(k, \vec{\Gamma}, E)$
    if $T \neq \emptyset$ **then**
       $index := Select(T, F)$
       forall the $\Gamma(i) \in T$ **do**
       | $\Gamma(i) = \Gamma(index)$
       end
    end
end
return $\vec{\Gamma}$

Algorithm 6.1: Basic ESR run as required for the application of the prediction model.

Assumptions. In the following, three assumptions are given that are required to be valid for the application of the prediction model. Their purpose is to provide a simplified view on an ESR run, and to still capture its essential properties. The assumptions may seem to be rather restrictive, but nevertheless they only require conditions that are valid most of the time in most ESR scenarios (cf. discussion below).

1. Assumption 1 – "Restriction to Tournament Selection": Reproduction is performed by a tournament selection-like procedure as described above.

2. Assumption 2 – "Twofoldness of the Population": At any point in time, the population $\vec{\Gamma}$ can be divided into two (possibly empty) sub-populations, each of which contains only individuals of nearly equal (i. e., indistinguishable by long-term observation) quality in terms of behavioral performance.

3. Assumption 3 – "Individual Fairness": On average and in the long run, the reproduction capability of a robot depends only on its own controller and the controllers of the other robots, but not on other properties of the environment. Particularly, a robot's long-term chance of being selected is independent of its current state in the environment. In other words, every robot should be capable of improving its explicit and implicit fitness solely by changing its controller.

Discussions of the three assumptions.

- Discussion of assumption 1 – "Restriction to Tournament Selection": Due to the existence of communication constraints (e. g., a limited communication distance or a limited number of communication channels), many selection methods have to be ruled out in a real-world scenario. Tournament selection is a slim selection method in terms of communication complexity which has proven successful in many classic EC scenarios as well as in ESR, cf. Chap. 4. Furthermore, natural selection can be seen as a complex generalization of tournament selection. While this selection method is not the only way to define the success prediction model, it seems to be the most natural of all the well established selection methods to start with.

- Discussion of assumption 2 – "Twofoldness of the Population": Every population $\vec{\Gamma}$ that includes at most two qualitatively different behaviors can obviously be split into a *superior* subpopulation $\mathcal{S} \subseteq \vec{\Gamma}$ and an *inferior* subpopulation $\mathcal{I} = \vec{\Gamma} \setminus \mathcal{S}$, each possibly empty. Assuming at most two different behaviors is less restrictive than it looks at first sight. In fact, it describes the most commonly observed situation during typical ESR runs. There, most of the time populations reside on neutral plateaus containing only minor mutations of the same behavior without significant influence on selection. Once in a while, a bigger mutation occurs leading to the situation of two qualitatively different behaviors in the population. At this point, the model's purpose is to predict the chances of the two behaviors, the superior and the inferior, to occupy the population – eventually leading to the next neutral plateau. For the most part (particularly in small populations), this is also given in natural evolution where genetic drift leads to large neutral plateaus of nearly equal individual properties [98]. From this point of view, the capability of repeatedly selecting superior over inferior individuals after behavior-changing mutations reflects the expected success of evolution. On the other hand, the case of more than two significantly different behaviors occurring in a population at the same time is rare and assumed to be negligible.

- Discussion of assumption 3 – "Individual Fairness": This assumption is a necessary precondition for most reasonable ESR scenarios. It states that on average every robot's chance of being implicitly selected should depend on its behavior only, which is just the claim that any robot should be able to control its selection probability by evolving its controller. For example, if robots reproduce when meeting each other, it has to be assured that no obstacle in the environment completely prohibits reproduction for parts of the population by separating them in a closed area. In a real-world scenario, robots can get trapped and be henceforth unable to improve their situations whatever their own controllers or the controllers of the other robots evolve to. However, this seems to be a hardware or environmental problem rather than a problem of the evolution process. This particular special case can be included in the model by manually excluding the trapped robot from further calculations. Assumption 3 is owed to the application perspective of ESR and the (unspoken) claim to exploit available robot hardware as far as possible. Conversely, an analogy to sickness or death in nature, which could in principle be modeled by affecting the functioning of the actual robot devices, does not seem to be of any practical use for the model.

Relative behavioral quality of a population. The relative behavioral quality of a population $\vec{\Gamma}$ with respect to a subpopulation $P \subseteq \vec{\Gamma}$ is defined as $R(\vec{\Gamma}, P) =_{def} a/b$ with $a = |P|$ and $b = |\vec{\Gamma}| - a$ being the number of individuals in subpopulation P and $\vec{\Gamma}$ without P, respectively. In the following, the subpopulations are assumed to be $P = \mathcal{S}$ and $\vec{\Gamma} \backslash P = \mathcal{I}$, and the shorter denotation $R(\vec{\Gamma}) = a = |\mathcal{S}|$ is used if the meaning is obvious from context. In other words, the relative quality of a population is given by the number of its superior individuals.

Markov chains. Given a finite state space M, a (first order) Markov chain is given by the probabilities to get from one state to another. There, the Markov property has to be fulfilled, which requires that future states depend only on the current state, but not on the past states. A Markov chain is defined in a common way according to the literature [58].

Definition 6.1

Markov chain and its basic properties.

A (homogeneous) Markov chain *with* $n \in \mathbb{N}$ *states is given by the state space* $M = \{m_1, \dots, m_n\}$ *and a matrix of probabilities* p_{ij}, $1 \leq i, j \leq n$ *determining the probability to get in one step from* m_i *to* m_j. M *is called* transition matrix.

> A state m_i is called absorbing *if and only if* $p_{ii} = 1$ *and for all* $j \neq i : p_{ij} = 0$. *A Markov chain is called* absorbing *if it includes at least one absorbing state and every non-absorbing state can reach an absorbing state (possibly in more than one step). All non-absorbing states of an absorbing Markov chain are called* transient.

6.2 A Model Capturing Completely Implicit Selection

In this section, as a first step toward a general success prediction model, a *Completely Implicit Selection (CIS)* is assumed to be underlying the evolution process. The CIS prediction model works without any explicit fitness function, i. e., the function f in Alg. 6.1 is assumed to return the constant value 1.

6.2.1 Two Parents per Reproduction (CIS-2)

The simplest version of the CIS model works with two parents for reproduction which is given by tournament size $k = 2$. This model is called CIS-2 in the following, and it captures a scenario where a robot has to "find" (whatever this means in a particular environment) exactly one other robot for reproduction. The explicit part of reproduction within a given tournament is then performed by uniformly randomly selecting one robot to copy its controller to the other robot. Therefore, the environment implicitly selects for the ability of finding another robot, but explicit selection within a tournament is turned off. Stated in the terminology introduced above, individuals capable of approaching other robots are superior over others that do not have this capability. Such superior individuals are favored implicitly, but not explicitly over inferior individuals, creating a selection pressure toward the implicit goal. This selection pressure is low, however, as a robot can wait for another robot to find it without reducing its chances of winning the reproduction tournament. There, the question arises, how much selection pressure exactly can be maintained in this extreme scenario.

In the following it is shown that in the CIS-2 model (and in any CIS-k model with $k \in \mathbb{N}$, for that matter), despite the existence of an implicit selection pressure toward superior individuals, long-term chances of surviving in a population are equal for inferior and superior individuals. This leads to populations that temporarily may learn to match the implicit selection criterion, but eventually have to fall back to random behavior. It is, however, important to note that this does not mean that a behavior that matches the implicit selection criterion can only occur by pure chance. Rather, implicit selection inevitably has to yield such behaviors once in a while, where the exact chance depends on the probability to bootstrap from trivial behavior. It only means that this behavior, once occurred, has the same probability of persisting in the population as any other behavior. Consequently, an oscillation between improvement and

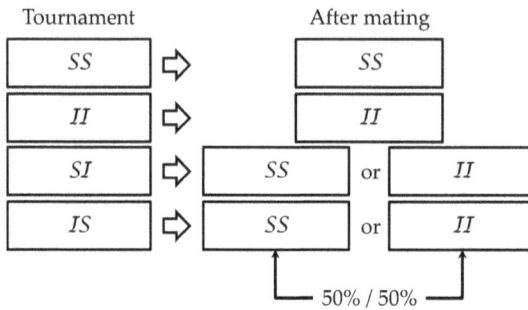

Fig. 6.2. The four possible situations in a reproduction tournament of the CIS-2 model. Without explicit fitness, the winner of a tournament is drawn uniformly randomly. Therefore, chances are equal for the population to gain or to lose a superior individual in the SJ and JS cases.

random behavior has been observed in the simulation experiments, cf. Sec. 6.4.2. The implications of this result for applications, and the relevance of the CIS model for the theoretical understanding of evolution are discussed at a higher level in Sec. 6.4.3.

In the general CIS-2 scenario, two robots reproduce when they match some arbitrary environmental criterion, and the tournament winner is chosen by uniform distribution. As each of the robots in a tournament may be from one of the sets S or J, one of the following four situations can occur within a mating tournament ("XY" meaning that one individual is from set X and the other from set Y): JJ, JS, SJ or SS. The cases JJ and SS, cause no changes to the population; for the cases JS and SJ chances are 0.5 for both overwriting the controller from S with the controller from J and vice versa. Therefore, the population either gains a new S robot and loses an J robot or the other way around, cf. Fig. 6.2.

The reproduction process of a population with n individuals can be expressed as an $n + 1 \times n + 1$ Markov matrix reflecting the transition probabilities between the $n + 1$ possible population states given by the relative behavioral quality of a population. The rows and columns correspond to population states, and each matrix entry denotes the probability of changing from one state to another. The $n + 1$ states are denoted by $0/n$, $1/n\text{-}1$, ..., $n/0$ where a/b means a superior and $b = n - a$ inferior robots in the population. An entry p_{ij} equals the probability that a population that is currently in state i changes to state j after one mating event. This implies that every row of the matrix sums up to 1. For the mating procedure described above, the matrix P_{CIS-2} is given by

$$P_{CIS-2} =$$

$$
\begin{array}{c}
\\
^0/_n \\ \vdots \\ ^i/_{n-i} \\ \vdots \\ ^n/_0
\end{array}
\begin{array}{ccccccc}
^0/_n & \cdots & ^{i-1}/_{n-i+1} & ^i/_{n-i} & ^{i+1}/_{n-i-1} & \cdots & ^n/_0 \\
\left(\begin{array}{ccccccc}
1 & & & & & & \\
& \ddots & & & & 0 & \\
& & \frac{c_i}{2} & s_i & \frac{c_i}{2} & & \\
& 0 & & & & \ddots & \\
& & & & & & 1
\end{array}\right)
\end{array}
$$

with

$$\forall i \in \{1, \ldots, n-1\} : c_i, s_i \in [0, 1], \; c_i + s_i = 1.$$

There, c_i, s_i are the probabilities that, in a population state $^i/_{n-i}$, a reproduction induces a state change (c_i; when two different robots reproduce) or that the population stays in the same state (s_i; when two uniform robots reproduce), respectively. In the states $^0/_n$ and $^n/_0$ there are no different individuals in the population, therefore, no state change can be induced by reproduction. These states are absorbing, and if one of them is entered, the population remains stable henceforth.

6.2.2 Eventually Stable States (k=2)

The transition matrix P_{CIS-2} defines a homogeneous Markov chain with the state set $S =_{def} \{^0/_n, ^1/_{n-1}, \ldots, ^n/_0\}$ and the set of absorbing states $A =_{def} \{^n/_0, ^0/_n\} \subseteq S$. As both absorbing states can be reached from any of the non-absorbing states $t \in S\backslash A$, the matrix P is absorbing; the transient states are denoted as $T =_{def} S\backslash A$.

For the purpose of success prediction, the long-term development of a population is sought which is reflected by an answer to the question if the population will eventually enter the desired stable state $^n/_0$ or the unwanted stable state $^0/_n$. In the former case, all robots finally receive a superior controller meaning that evolution has successfully selected the desired individuals. In the latter case, only inferior individuals are left in the population meaning that the selection mechanism has been incapable of preserving superior controllers. Therefore, the probability for eventually entering the stable state $^n/_0$ or $^0/_n$, respectively, is an indicator for the quality of the chosen selection mechanism.

If a transition matrix is raised to the power of n, an entry p_{ij} of the resulting matrix displays the probability that state j is reached after n steps when starting in state i. To receive the probabilities for the stable states to eventually get entered, the matrix has to be raised to infinity meaning that the limit

$$P^\infty_{CIS-2} =_{def} \lim_{n\to\infty} P^n_{CIS-2}$$

has to be calculated. The matrix P^∞_{CIS-2} exists for every absorbing Markov chain [58], meaning that every entry $p_{ij}^{(\infty)}$ converges. The limit matrix has non-zero entries in the columns which denote the absorbing states and zero entries at all other positions as every transient state has to be left for an absorbing state eventually. Therefore, it is only necessary to calculate the absorbing columns of the limit matrix.

For any absorbing transition matrix P the non-zero columns of P^∞ can be calculated by the following procedure. First, the *canonical form* CF_P of the matrix P is generated by shifting all absorbing states to the end in rows and columns such that an identity sub-matrix is built at the bottom-right corner of P. For the matrix P_{CIS-2}, the $\%_n$-state is already at the correct position; the $^n/_0$-state has to be shifted to the next to last position in rows and columns:

$$
CF_{P_{CIS-2}} = \left(
\begin{array}{ccccccc|cc}
s_1 & c_1 & \cdots & & & & & c_1 & 0 \\
 & & \ddots & & & & & 0 & 0 \\
\cdots & c_j & s_j & c_j & & \cdots & & \vdots & \vdots \\
 & & & & \ddots & & & 0 & 0 \\
 & & \cdots & c_{n-1} & s_{n-1} & & & 0 & c_{n-1} \\
\hline
0 & & \cdots & & & 0 & & 1 & 0 \\
0 & & \cdots & & & 0 & & 0 & 1
\end{array}
\right)
$$

The new matrix is now generally of the form

$$
CF_P = \left(\begin{array}{c|c} Q & R \\ \hline 0 & I \end{array} \right)
$$

where Q consists of transitions between transient states, R consists of transitions from transient states to absorbing states, I is an identity matrix reflecting transitions within absorbing states and 0 is a zero matrix. The matrix N_P with

$$
N_P =_{def} (I' - Q)^{-1}
$$

(where I' is an identity matrix with the same size as Q) is called the *fundamental matrix* of P. Now, in the matrix

$$
L_P =_{def} N_P \cdot R
$$

an entry l_{ij} reflects the probability that the absorbing chain will be absorbed in the absorbing state j if the process starts in state i. Therefore, L contains exactly the non-zero columns of the desired limit matrix P^∞. For the matrix P_{CIS-2}, independently of the c_i and s_i probabilities, the limit $L_{P_{CIS-2}}$ calculates to

$$
L_{P_{CIS-2}} =
\begin{array}{c}
 \\
{}^{0}/_{n} \\
{}^{1}/_{n-1} \\
{}^{2}/_{n-2} \\
\vdots \\
{}^{n-2}/_{2} \\
{}^{n-1}/_{1} \\
{}^{n}/_{0}
\end{array}
\begin{array}{cc}
{}^{0}/_{n} & {}^{n}/_{0} \\
\left(\begin{array}{cc}
1 & 0 \\
1-\frac{1}{n} & \frac{1}{n} \\
1-\frac{2}{n} & \frac{2}{n} \\
\vdots & \vdots \\
\frac{2}{n} & 1-\frac{2}{n} \\
\frac{1}{n} & 1-\frac{1}{n} \\
0 & 1
\end{array} \right)
\end{array}
$$

Therefore, the probability for ending in the superior state increases linearly with the number of superior robots at the beginning. Simultaneously, the probability for ending in the inferior state decreases at the same range. This result may be unintuitive due to the existence of implicit selection, but it is not unexpected as in the CIS-2 scenario the winner within a tournament is drawn by uniform probabilities. The order in which the superior or inferior robots are put together to tournaments does not affect the overall probabilities of reaching one of the stable states. The matrix $L_{P_{CIS-2}}$ can be visualized as shown by the dashed line in Fig. 6.5 of the next section, indicating a linear increase of the probability of converging to the superior stable state as a function of the number of superior individuals in the initial population.

This result further implies that the long-term success of CIS-2 scenarios is compromised, as during evolution, a population does not persist in a stable state, but it is "attacked" by mutations. They cause transitions from stable states to transient states (outside of the Markov process) and eventually back to another stable state (within the Markov process, assuming sufficiently long periods without behavior-changing mutations). If the chances for an inferior mutation to overrule the population are the same as for a superior mutation, the population cannot constantly remain in an improvement process. Usually, the initial population has a low behavioral quality, therefore, a CIS-2 scenario can lead to improvements in the beginning and repeatedly during the process of evolution, but they cannot remain stable in the long-term. The same is true for the generalized CIS-k scenario as is shown in the next section.

6.2.3 Tournament size k

The CIS-2 scenario can be stated in a more general form, called CIS-k, for a reproduction neighborhood of size $k \in \mathbb{N}$. Again, one of the k controllers is selected by a uniform probability to be copied to all other robots in a tournament, as explicit selection is still deactivated. Using the same notation as in Sec. 6.2.1, the CIS-k transition matrix P_{CIS-k} is given in Fig. 6.3. Fig. 6.3 As in the CIS-2 case, an entry p_{ij} of the matrix P_{CIS-k} reflects the probability that a population in state i switches to state j by a single mating event. Variables c_{ab} denote the probability that in a population that is currently

$$
P_{CIS-k} =
\begin{array}{c}
 \\ 0/n \\ 1/n\text{-}1 \\ 2/n\text{-}2 \\ \cdots \\ i/n\text{-}i \\ \cdots
\end{array}
\begin{array}{ccccccccccccc}
0/n & 1/n\text{-}1 & 2/n\text{-}2 & \cdots & k/n\text{-}k & k\text{+}1/n\text{-}k\text{-}1 & \cdots & i\text{-}2/n\text{-}i\text{+}2 & i\text{-}1/n\text{-}i\text{+}1 & i/n\text{-}i & i\text{+}1/n\text{-}i\text{-}1 & i\text{+}2/n\text{-}i\text{-}2 & \cdots \\
\boxed{1} & 0 & 0 & & 0 & 0 & & 0 & 0 & 0 & 0 & 0 & \\
\frac{k-1}{k}c_{1,1} & \boxed{c_{1,0}+c_{1,k}} & 0 & \cdots & \frac{1}{k}c_{1,1} & 0 & \cdots & 0 & 0 & 0 & 0 & 0 & \cdots \\
\frac{k-2}{k}c_{2,2} & \frac{k-1}{k}c_{2,1} & \boxed{c_{2,0}+c_{2,k}} & & \frac{2}{k}c_{2,2} & \frac{1}{k}c_{2,1} & & 0 & 0 & 0 & 0 & 0 & \\
 & & & & & & \ddots & & & & & & \\
0 & 0 & 0 & & 0 & 0 & \cdots & \frac{k-2}{k}c_{i,2} & \frac{k-1}{k}c_{i,1} & \boxed{c_{i,0}+c_{i,k}} & \frac{k-1}{k}c_{i,k-1} & \frac{k-2}{k}c_{i,k-2} & \cdots \\
 & & & \cdots & & & & & & & \ddots & &
\end{array}
$$

Fig. 6.3. General form of the transition matrix for the CIS scenario with k parents. The matrix reflects selection without an explicit fitness function in a population of size n, using mating tournaments of size k. Diagonal elements are marked by a surrounding box; they represent transitions where no state change occurs, i. e., when only one agent type (superior or inferior) is selected in a tournament. The matrix has at most $k - 1$ non-zero elements at the left and the right side of the diagonal elements of every row; the number decreases to 0 at the top and the bottom of the matrix. All rows sum up to 1.

in state $a/_{n\text{-}a}$ the next mating event is based on a $b/_{k\text{-}b}$ tournament, i. e., a tournament with b superior and $k - b$ inferior individuals. The diagonal elements of the matrix (marked by a box in the figure) denote the probability that the population state does not change by mating. Therefore, they are given by the sum of the probabilities for a tournament with only superior and a tournament with only inferior individuals, i. e., $p_{ii} = c_{i0} + c_{ik}$ (in the CIS-2 case this probability has been denoted by s_i). It is important to note that the values of the variables c_{ab} are parameters to the system that have to be obtained by observation in simulation runs or laboratory experiments; they build an interface to the real world allowing for experimental data to affect the prediction outcome. (In most of the experiments at the end of this chapter, the c_{ab} values are determined beforehand for testing reasons, but this is not the usual use case of the model.)

For calculating a bottom/left non-diagonal (non-zero) entry p_{ij} $(i > j)$, the probability variable $c_{i,i-j}$ has to be considered which reflects an $i\text{-}j/_{k\text{-}i+j}$ tournament occurring in an $i/_{n\text{-}i}$ population; such a tournament can turn a population from state $i/_{n\text{-}i}$ to $i/_{n\text{-}j}$. This probability has to be multiplied by the probability that an inferior individual will win the tournament (since $j < i$ means that the number of superior individuals decreases). As the individuals are drawn by uniform distribution from the tournament, this probability is depending only on the number of superior and inferior individuals given by $\frac{k-(i-j)}{k}$. The overall probability that defines an entry $p_{ij}, i > j$ is, therefore, given by

$$p_{ij} =_{def} \frac{k - (i - j)}{k} \cdot c_{i,i-j}$$

Analogously the top/right non-diagonal (non-zero) entries p_{ij}, $(i < j)$ can be computed to

$$p_{ij} =_{def} \frac{k - (j - i)}{k} \cdot c_{i,k-j+i}$$

For all c_{ij} in the matrix P_{CIS-k} it has to hold that

$$\forall i \in \{1, \ldots, n - 1\}: \sum_{j=\max(0,i+k-n)}^{\min(k,i)} c_{ij} = 1 \ \left(c_{ij} \in [0, 1]\right).$$

This implies that the sum of every row i of the matrix P_{CIS-k} is 1. During one reproduction, at most $k - 1$ individuals can be turned from S to J or vice versa. That is reflected by the fact that all probabilities c_{ij} with $j < 0$ or $j > k$ have to be zero. Therefore, at most the $k - 1$ elements left and right of the diagonal elements in matrix P_{CIS-k} are non-zero. Furthermore, in populations with $i < k$ superior $(n - i < k$ inferior) individuals, all probabilities c_{ij} with $j > i$ $(j > n - i)$ have to be zero as at most i superior $(n - i$ inferior) individuals exist in the population and, therefore, can be in a tournament. Accordingly, the matrix C which is given by the probabilities c_{ij} for $0 \le i \le n$ and

$0 \le j \le k$ has the form

$$
C = \begin{array}{c} \\ ^0/_n \\ ^1/_{n\text{-}1} \\ \vdots \\ ^i/_{n\text{-}i} \\ \vdots \\ ^{n\text{-}1}/_1 \\ ^n/_0 \end{array}
\begin{array}{ccccccc}
^0/_k & ^1/_{k\text{-}1} & ^2/_{k\text{-}2} & \cdots & ^{k\text{-}1}/_1 & ^k/_0 \\
\left(\begin{array}{cccccc}
1 & 0 & 0 & \cdots & 0 & 0 \\
c_{1,0} & c_{1,1} & 0 & \cdots & 0 & 0 \\
\vdots & \vdots & & & \vdots & \\
c_{i,0} & c_{i,1} & c_{i,2} & \cdots & c_{i,k-1} & c_{i,k} \\
\vdots & \vdots & & & \vdots & \\
0 & 0 & 0 & \cdots & c_{n-1,k-1} & c_{n-1,k} \\
0 & 0 & 0 & \cdots & 0 & 1
\end{array}\right)
\end{array}
$$

There are $k \cdot (n - k) + n + 1$ non-zero entries in C. As stated above, every row in C has to sum up to 1 to assure that every row in P_{CIS-k} sums up to 1.

6.2.4 Eventually Stable States (arbitrary k)

By the same procedure as in Sec. 6.2.2, the probabilities for a population to eventually reach the stable states $^0/_n$ and $^n/_0$ when starting in a state $^i/_{n\text{-}i}$ can be computed. As the choice of an individual in a tournament is still uniform, for all tournament sizes k and all probability matrices C the probability distribution is the same as in the CIS-2 case:

$$
L_{P_{CIS-k}} = L_{P_{CIS-2}}
$$

However, the expected time to absorption, i. e., the number of mating events until a stable state is reached, decreases if k is increased. The expected time to absorption of a Markov chain given by a matrix P is given by a column vector t whose entries t_i reflect the expected number of state changes until the chain reaches an absorbing state if starting in state i. The vector t can be computed as

$$
t = N_P \cdot v
$$

where N_P is the fundamental matrix of P (cf. Sec. 6.2.2) and v is a column vector all of whose entries are 1. For example, the expected time to absorption for population size $n = 10$ and tournament sizes $k = 2, \ldots, 9$ is depicted in Fig. 6.4. Obviously, the time to absorption decreases drastically from tournament size 2 to tournament sizes 3 and 4. However, it is important to note that "time" is measured here in terms of the number of reproduction events. Depending on the environment it may take longer in terms of evolution time to select tournaments of bigger size than those of smaller size.

6.3 Extending the CIS Model to Capture Explicit Selection

Explicit fitness can be introduced into the purely implicit selection model CIS by making the probability for superior individuals to be winners in a tournament different

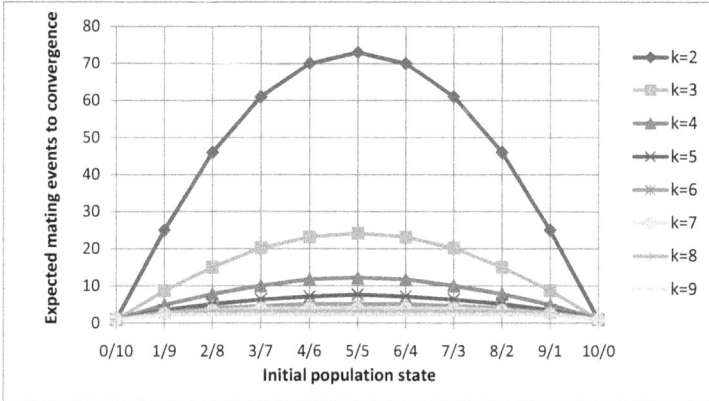

Fig. 6.4. Expected absorption times for different tournament sizes. The time to absorption is given in terms of reproduction events as a function of the initial population state $(0/10, \ldots, 10/0)$ for tournament sizes $k = 2, \ldots, 9$ in a population with $n = 10$ individuals.

(higher, in the regular case) from that of inferior individuals. The model described in this section is called *Explicit and Implicit Selection (EIS)* model. In the evolution process described by Alg. 6.1, explicit fitness is given by the function f. It is calculated from environmental variables and is intended to measure behavioral qualities with respect to the desired behavior. In ESR, factors such as noise in the environment, delayed fitness calculation and erroneous design of the fitness function can corrupt the fitness measure. Therefore, the probability that a superior individual is selected explicitly over an inferior individual is usually below 1, even if there is no probabilistic algorithm involved in the selection process.

To reflect the influence of f on selection, a *confidence factor* $c \in [0, 1]$ is introduced which states how accurately f differentiates between superior and inferior individuals. A zero value for c means that explicit fitness cannot increase the chance that a superior individual is chosen in a tournament; in this case, the EIS model is equivalent to the CIS model. A high c value, on the other hand, means that it is likely for a superior individual to be chosen over an inferior one (there, f expectedly returns higher values for superior than for inferior individuals); $c = 1$ implies that in every tournament that contains at least one superior individual such an individual will win. In the following, a below-zero value for c is not considered assuming that in any reasonable scenario selection can be expected on average to select a superior over an inferior individual.

The confidence factor is included into the model by extending the matrix P_{CIS-k}. At the bottom/left side of the diagonal of the transition matrix the entries are multiplied with $(1 - c)$. At the top/right side of the diagonal the enumerator $k - (j - i)$ is replaced with $k - (j - i)(1 - c)$. In this way the chance for switching to a state right of the diagonal gets higher when c is increased (to maximally the $c_{i,i-j}$ value) while the chance for switching to a state left of the diagonal decreases (to minimally 0); for $c = 0$ nothing changes. The diagonal entries do not have to be changed as the corresponding tournaments consist of uniform individuals. It can easily be seen that each row of the matrix still sums up to 1.

Generally, for population size n, tournament size k and fitness confidence factor c the $(n + 1) \times (n + 1)$ transition matrix P_{EIS-k} is given by its entries

$$
p_{ij} = \begin{cases}
c_{i0} + c_{ik} & \text{if } i = j, \\
\frac{(k-(i-j))(1-c)}{k} c_{i,i-j} & \text{if } i - k < j < i, \\
\frac{k-(j-i)(1-c)}{k} c_{i,k-j+i} & \text{if } i < j < i + k, \\
0 & \text{otherwise}
\end{cases}
$$

for $0 \le i, j \le n + 1$. A complete inner row of the most general form of the transition matrix is given in Tab. 6.1. The restrictions to the tournament probability values given in matrix C in Sec. 6.2.3 have to be respected here as well.

Table 6.1. Non-zero entries of an inner row i/n-i of a general EIS transition matrix. The first heading denotes the column of the transition matrix, the second the corresponding tournament, i. e., the column of the probability matrix C. For better readability, the row is divided into three parts, one showing cells left of the diagonal, another the cell at the diagonal and the last cells right of the diagonal.

Trans. Mat.	$i\text{-}k\text{+}1/_{n\text{-}i\text{+}k\text{-}1}$	$\cdots$	$i\text{-}2/_{n\text{-}i\text{+}2}$	$i\text{-}1/_{n\text{-}i\text{+}1}$	
Tournament	$k\text{-}1/_1$	$\cdots$	$2/_{k\text{-}2}$	$1/_{k\text{-}1}$	$\cdots$
$i/_{n\text{-}i}$	$\frac{1-c}{k} c_{i,k-1}$	$\cdots$	$\frac{(k-2)(1-c)}{k} c_{i,2}$	$\frac{(k-1)(1-c)}{k} c_{i,1}$	

$i/_{n\text{-}i}$	
$\cdots$ $k/_0$ or $0/_k$ $\cdots$	
$c_{i,k} + c_{i,0}$	

	$i\text{+}1/_{n\text{-}i\text{-}1}$	$i\text{+}2/_{n\text{-}i\text{-}2}$	$\cdots$	$i\text{+}k\text{-}1/_{n\text{-}i\text{-}k\text{+}1}$
$\cdots$	$k\text{-}1/_1$	$k\text{-}2/_2$	$\cdots$	$1/_{k\text{-}1}$
	$\frac{k-(1-c)}{k} c_{i,k-1}$	$\frac{k-2(1-c)}{k} c_{i,k-2}$	$\cdots$	$\frac{k-(k-1)(1-c)}{k} c_{i,1}$

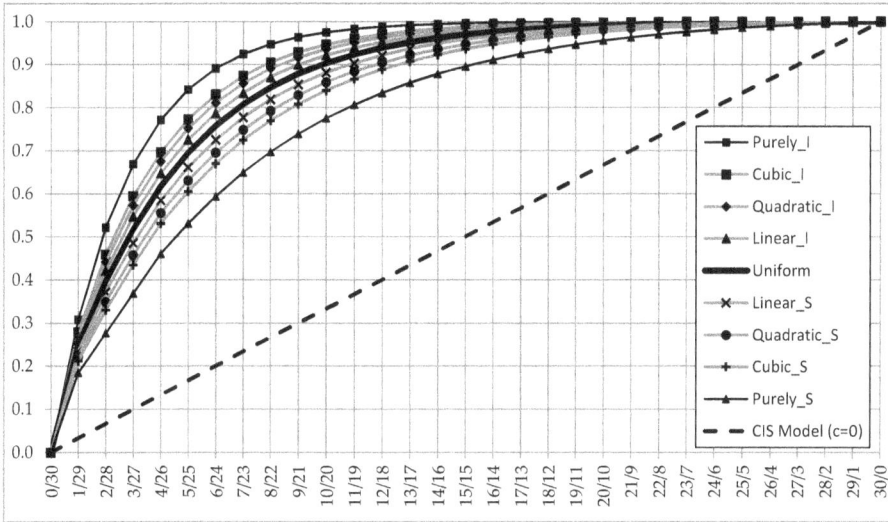

Fig. 6.5. Probabilities for eventually converging to the superior state n/0 in the EIS model. The X axis denotes the initial population states $i/n - i$, the Y axis denotes the probabilities for eventually converging to the superior stable state. The plot "Uniform" refers to a uniform distribution in C, the plots "Cubic_I", "Quadratic_I" and "Linear_I" refer to distributions with higher chances for inferior tournaments, the plots "Cubic_S", "Quadratic_S" and "Linear_S" to distributions with higher chances for superior tournaments. The plots "Purely_I" and "Purely_S" refer to the extreme cases where only 1/3 and 3/1 tournaments are selected, respectively (cf. probability matrices given in Fig. 6.6 and the detailed description in the text).

Convergence to a stable state and influence of the probability matrix C. In an EIS scenario, in contrast to the situation using the CIS model, the probabilities in C can have an impact on the convergence probabilities of an evolutionary run. The following purely theoretical result shows that in an EIS scenario the probability of selecting superior individuals can increase more than linearly with the number of superior individuals in the initial population, and that this increase is dependent of the matrix C. There, a population with $n = 30$ individuals, a tournament size of $k = 4$, and nine different probability matrices (given below) are assumed. Successful long-term selection is given by preferring superior over inferior individuals using a confidence factor of $c = 0.2$.

Fig. 6.5 shows the probabilities of a population for converging to the superior state $n/0$ as a function of the initial population state which corresponds to the number of superior individuals in the initial population. The thick black plot in the middle corresponds to a uniform distribution in every row of C. For the 3 gray plots right/down of it, the probability for a tournament $i/k-i$ to occur is increased in a linear, quadratic or cubic manner with the number of superior individuals (more precisely, the probability is set to $(i + 1)^e$ for $e \in \{1, 2, 3\}$ and then normalized such that every row of C sums up to 1; the population state is not taken into account, i. e., every column has the same value in every position). Symmetrically, for the 3 gray plots left/top of the middle, the probability is increased with the number of inferior individuals (i. e., for a tournament $i/_{k-i}$ it is set to $(k-i+1)^e$ and then normalized). The corresponding matrices are depicted graphically in Fig. 6.6.

The leftmost and rightmost plots labeled "Purely_I" and "Purely_S", respectively, belong to extreme settings where selection is performed with a probability of 1.0 in $1/_3$ and $3/_1$ tournaments, all other entries of C being set to 0 (except for the impossible cases in the three upper or lower rows; here the column which is as close as possible to the $1/_3$ or $3/_1$ tournament, respectively, is set to 1.0). These two plots can be seen as the limits of the polynomial plots described above for $e \to \infty$. In other words, all polynomial probability matrices of the above form result in plots within this range. Different from the CIS case, the number of individuals converted from inferior to superior is not symmetrical to the number converted from superior to inferior. Rather, the plots of the EIS model are left/top of the linear CIS plot, depicted by a dashed line in the figure for comparison, indicating a more than linearly increasing chance of selecting superior over inferior individuals. Generally it holds that the further to the left/top a plot is, the higher is the corresponding capability of the explicit fitness function to recognize superior individuals.

It is a rather unsurprising result that the EIS model is capable of modeling an increase in explicit selection pressure in the given setting, as it has been defined accordingly. But beyond that, it is interesting to note that the probability matrices that gain the highest chances for eventually converging to the superior state (leftmost plots) are those that provide the highest chance of implicitly selecting *inferior* tournaments. Accordingly, the matrices with the lowest success rate are those that implicitly select mainly for *superior* tournaments. This observation is quite contrary to a first intuition suggesting that selection should always (explicitly and implicitly) favor superior individuals over inferior ones in order to improve evolution. In the given setting, implicitly selecting tournaments with few superior individuals that, in consequence, have a relatively high chance of converting a lot of inferior individuals, pays off more than selecting superior tournaments which can only convert few inferior individuals at a time and include a risk that an inferior individual converts a lot of superior individuals. Conversely, the situation is much simpler for explicit selection where preferring superior over inferior individuals always pays off (in terms of the prediction model; nevertheless, the standard problem of an insufficient population diversity may require to moderate a "too acurate" explicit selection as well).

6.4 Experiments

In the first part of this section the evolutionary setup for the experiments is described. Afterwards, the experimental results are presented and discussed. The EIS model is

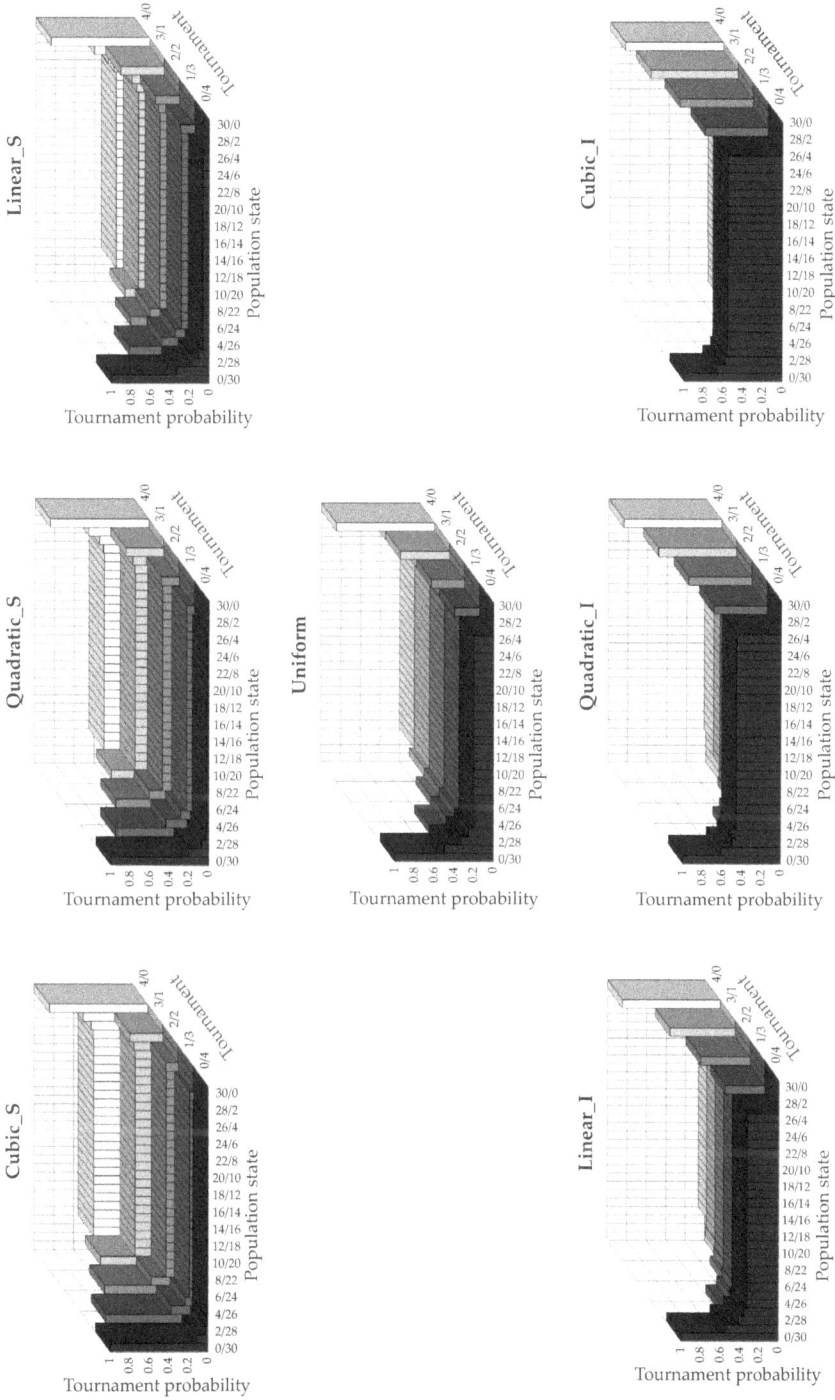

Fig. 6.6. Polynomial probability matrices used in the theoretical example. The matrices correspond to the accordingly labeled plots given in Fig. 6.5.

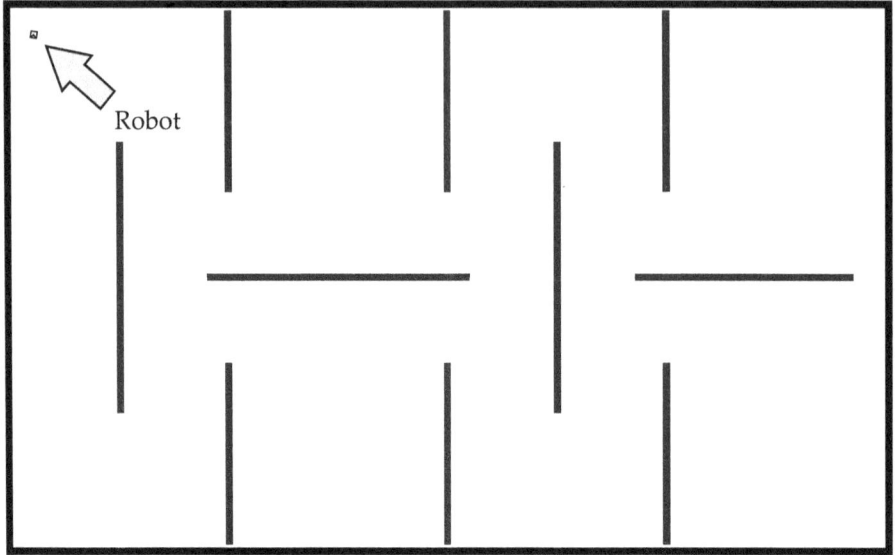

Fig. 6.7. The experimentation field with a robot drawn to scale. Walls are placed such that simply driving straight forward is ineffective; a Wall Following behavior is used for the superior individuals causing them to perform a combination of driving forward and turning behavior.

firstly applied to two rather artificial ESR scenarios with a centralized selection operator. Secondly, a more realistic decentralized ESR scenario is studied. Furthermore, experimental results with the CIS model are discussed briefly regarding their relation to the according model prediction as well as their real-world implications.

6.4.1 Evolutionary setup

All experiments are performed in simulation and use the basic evolutionary framework introduced in Chap. 4 as an application of Alg. 6.1. For most experiments mutation is switched off, and two hard-coded behaviors, different in terms of "superiority", are spread over the population (note that "superiority" in the above-defined sense is correlated but not equal to explicit fitness expectation).

The experimentation field is given in Fig. 6.7. In all experiments the populations consist of $n = 30$ robots which are placed at random positions in the field facing at random directions (all random operations are performed by a uniform distribution over the respective range). The desired behavior in terms of "superiority" is given by the capability of driving as far as possible in a constant time period. Accordingly, the explicit fitness snapshot (cf. Chap. 4) is calculated by summing up every 10 simulation steps the distance driven since the last snapshot. There, Euclidean distance is used meaning that driving curves is less effective than driving straight forward. Addition-ally, the current fitness value is divided by 1.3 after each snapshot due to evaporation (cf. Chap. 4). Except for the experiment concerning the CIS model, all evolutionary runs are performed without mutation, and until convergence to a stable population state (i. e., $n/0$ or $0/n$).

6.4.2 Experimental Results Using the EIS Model

The first part of the experiments is intended to study the EIS model's capability of correctly predicting the probabilities of a population converging to the superior stable state $n/0$ or the inferior stable state $0/n$, respectively. As the convergence to the superior state correlates to a proper selection mechanism, the relation of runs ending up in the superior state is a measure of the system's success rate. Therefore, this part of the experiments is actually intended to study if the EIS model is capable of correctly predicting an evolutionary system's success rate.

The initial populations are divided into two sets S and J of individuals which both perform suboptimal behaviors in terms of implicit and explicit desirability. There, individuals from S perform a "superior" and individuals from J an "inferior" behav-ior. The superior individuals are equipped with a Wall Following behavior letting the robots effectively explore the field. The inferior individuals are constantly driving small circles by switching in every step between a driving and a turning state and, therefore, they are expected to have a lower fitness than the superior individuals (al-though, due to environmental properties, this is not necessarily always the case). The initial population state is varied within the state space $S' = \{1/29, 2/28, \ldots, 29/1\}$ (the states $0/30$ and $30/0$ are already converged, therefore, they are not tested).

First, a global selection operator based on a fixed probability matrix C is as-sumed to select the mating tournaments (cf. Sec. 6.2.3), i. e., if a population is in state i/n-i, row i of matrix C is used to determine the probabilities for the tournament types $(0/k, \ldots, k/0)$ to select. According to these probabilities, a tournament type i/k-j is selected and such a tournament is chosen uniformly randomly from the current population for mating. Afterwards, the tournament winner is chosen according to the explicit fitness as described in Alg. 6.1. As the quality of the fitness function is not known in advance, the confidence factor c is unknown. The aim of the first experi-ments is to show that there exists a confidence factor c such that the experimental data matches the model prediction. Originally, another intention has been to use this experimental setup as a preliminary experimentation framework for real-world sce-narios, allowing to find a constant c that can be used for a range of settings without having to repeatedly re-estimate it. However, as shown below, this simple approach is not practicable as c is not independent of the probability matrix C, i. e., a real-world scenario with its own probability matrix C requires a new estimation of c as well.

Fig. 6.8 shows the results of an experiment where the probabilities of matrix C

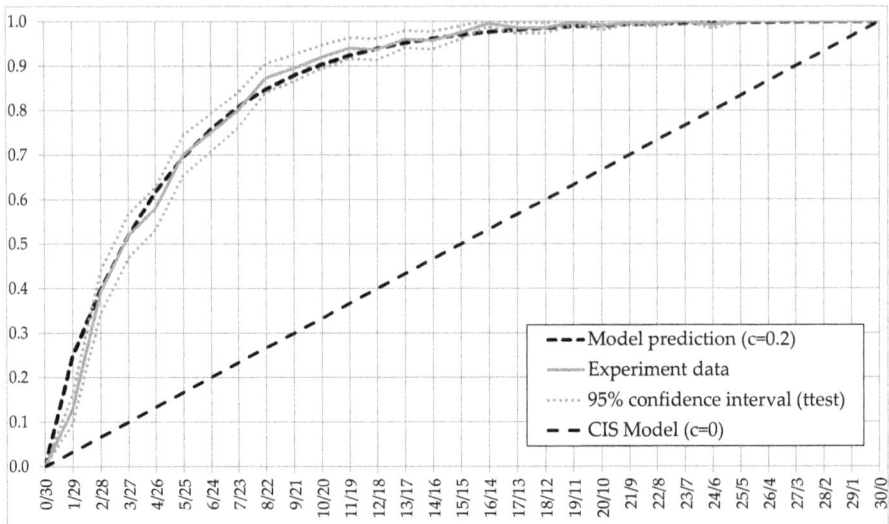

Fig. 6.8. Probabilities for eventually converging to the superior state 30/0 (uniform selection).
Matrix C is set to a uniform distribution in every row, tournament size to $k = 4$. The gray solid line
shows the average values from the experimental data, the gray dotted lines denote the according
95 % confidence interval. The thick black dashed line shows the model prediction with $c = 0.2$. The
thin black dashed line corresponds to the CIS model prediction.

are uniformly distributed within every row, i. e., every tournament formation has an
equal selection probability within the same row of C (except for impossible formations
which are set to 0). The matrix with tournament size $k = 4$ is depicted in the middle
part of Fig. 6.6, labeled "Uniform". The experiment is run 400 times for each of the
non-stable initial population states $1/29, \ldots, 29/1$. The chart shows the probabilities
of converging to the superior state $30/0$ as a function of the initial population state.
The percentage of experimental runs that converged to the superior state are shown
by the gray solid line. Two gray dotted lines denote the according 95 % confidence
interval given by a Student's ttest calculation. The thick black dashed line denotes a
prediction by the EIS model using a confidence factor of $c = 0.2$. This confidence factor
has been determined by a minimal error calculation in steps of 0.1. I. e., all model data
points are subtracted from the according experiment data points, adding the absolute
values to an error sum; then the value of $c \in \{0.0, 0.1, \ldots, 1.0\}$ is determined for
which the error sum is minimal. For comparison, the thin dashed black line shows
the CIS model prediction. As the experimental data points follow the model prediction
in practically all cases within the 95 % confidence interval (with the exception of the
initial population states $1/29$, $16/14$ and by a very minor error $19/11$), it can be safely
concluded that the model prediction is accurate in this experiment (with respect to
the "guessed" c value).

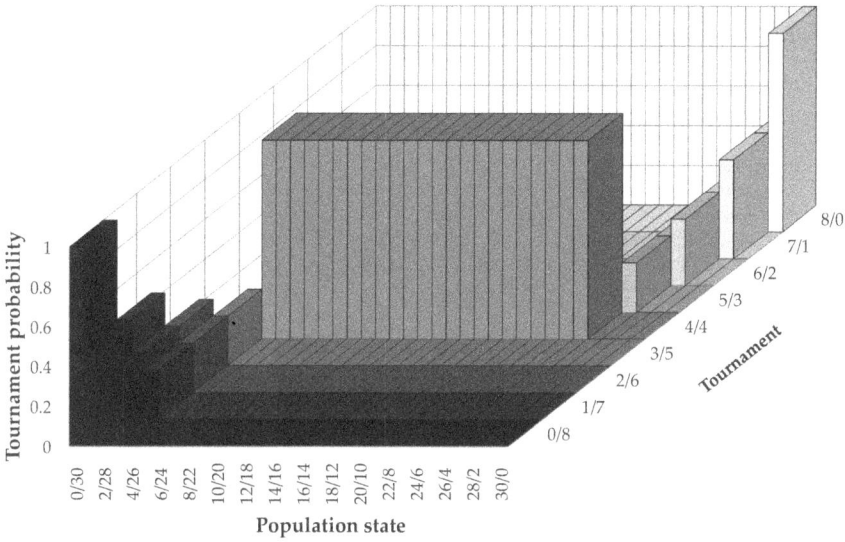

Fig. 6.9. Special probability matrix causing "jumps" in the resulting plots. Using a tournament size of $k = 8$, the matrix strongly favors $4/4$ tournaments. As a result, an increased number of superior individuals can lead to a decreased chance of converging to the superior stable state, cf. Fig. 6.10.

A second experiment has been performed with a tournament size of $k = 8$ by using a special matrix C which is highly unrealistic for any real scenario. There, C has a high probability of selecting a $4/4$ tournament, and a low (virtually zero) probability of selecting all the other tournaments. To be exact, the probability for selecting a $4/4$ tournament is set to $1 - 10^{-4}$ while all the other possibilities uniformly divide the remaining value of 10^{-4} among them. In rows where the $4/4$ tournament is not applicable, the other tournaments are uniformly distributed. Fig. 6.9 depicts this probability matrix. Due to the symmetry of the preferred tournament $4/4$, a counterintuitive property arises: the model predicts that there should be "jumps" in the probability plot, i. e., an *increase* in the number of superior individuals in the initial population can cause a *decrease* of the probability of converging to a superior state. This is, e. g., the case at the initial population states $5/25$ and $10/20$, cf. Fig. 6.10.

This experiment has been repeated 1,000 times for each of the 29 non-stable initial population states. The chart in Fig. 6.10 shows the according experimental results as well as the model prediction. Again, the best-fitting confidence factor has been calculated to be $c = 0.2$. For most of the data points the model lies within the 95 % confidence interval which, however, is tighter than in the above experiment due to the increased number of simulation runs. Additionally, and more importantly, the jumps predicted by the model occur in the experimental data as well. This particular observation has been interpreted as a strong indicator that the model prediction is prin-

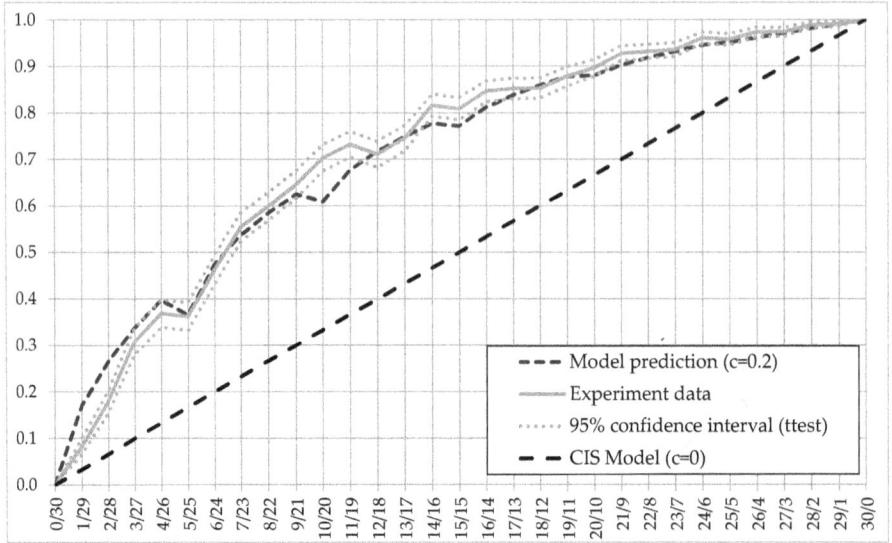

Fig. 6.10. Probabilities of converging to the superior stable state (cf. above chart). Tournament size is set to $k = 8$, matrix C is set to favor $4/4$ tournaments. As above, the gray lines denote experiment data and the black dashed line is the model prediction with $c = 0.2$. For comparison the CIS model prediction is plotted in this chart as well.

cipally correct. However, the second jump does not occur at state $10/20$ as predicted, but at state $12/18$; the other jumps seem to be at the correct positions, although they get smaller with increasing number of superior individuals in the population. While it is possible that this discrepancy is due to statistical errors, this explanation is rather unlikely as the two sequent values for the states $10/20$ and $11/19$ are considerably outside the 95 % confidence interval. If the inconsistency remains in future experiments, this particular special case can be utilized as a starting point for further improving the model.

A third experiment without mutation has been performed in a more realistic scenario using a decentralized selection method. Here, robots are selected for a mating tournament if they come spatially close to each other rather than by a global selection operator. A tournament winner is chosen as before, based on the explicit fitness values as described in Alg. 6.1. After having been part of a tournament, the robots are excluded from selection for 50 simulation steps to allow for an accurate fitness measurement with a possibly new controller. The radius for reproduction has been set to 210 mm which is, in relation to the field size, big enough to assure that nearly all runs converge eventually. For the small percentage of runs that have not converged within the first 200,000 simulation steps (around 1 % of all runs) the runs have been terminated and counted as "converged to superior" if the number of superior individuals

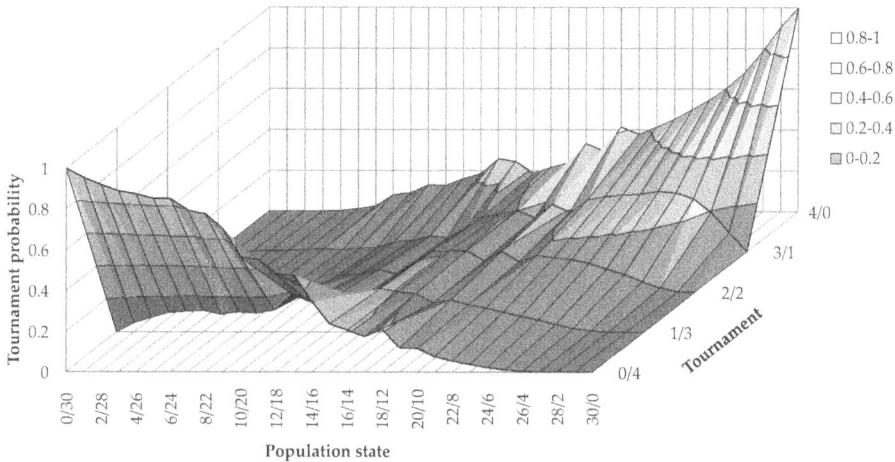

Fig. 6.11. Probability matrix originating from a decentralized reproduction strategy. The matrix is calculated as an average over all tournaments occurred in all simulation runs, and used as an input for the EIS model prediction given in Fig. 6.12.

in the last population has been at least 15 and as "converged to inferior" in all other cases. The tournament size has been set to $k = 4$, and the runs have been repeated 400 times for each of the non-stable initial population states.

In this experiment, tournaments have been selected in a decentralized way, therefore, there is no predefined probability matrix C. Rather, C is given by the environment and the selection parameters, and can be measured during a run. The chart in Fig. 6.11 depicts the probabilities of the matrix found by averaging over all occurred tournaments during all the $400 \cdot 29 = 11{,}600$ runs of this experimental setup. It can be observed that for a rather heterogeneous population with an approximately equal number of superior and inferior individuals the tournament probabilities are roughly uniform. With more superior individuals in the population, the probability for superior tournaments grows, and the other way around. This seems to be quite intuitive and it can be suspected that similar decentralized selection methods yield similar probability distributions. Therefore, this matrix or another matrix obtained from observation of a generalizable experimental setup can be chosen for success prediction of scenarios where the matrix C is unknown in advance.

Fig. 6.12 shows the convergence probabilities resulting from this experiment. This time the EIS model prediction has been calculated using the measured values for matrix C as depicted in Fig. 6.11. The confidence factor c is set to 0.06 which is the best approximation by a precision of 0.01. Again, for nearly all data points the model prediction is within the 95 % confidence interval. Obviously, the chance of converging to the superior state is raised by the explicit fitness function. However, the confidence factor c is decreased considerably compared to the above experiments with a global

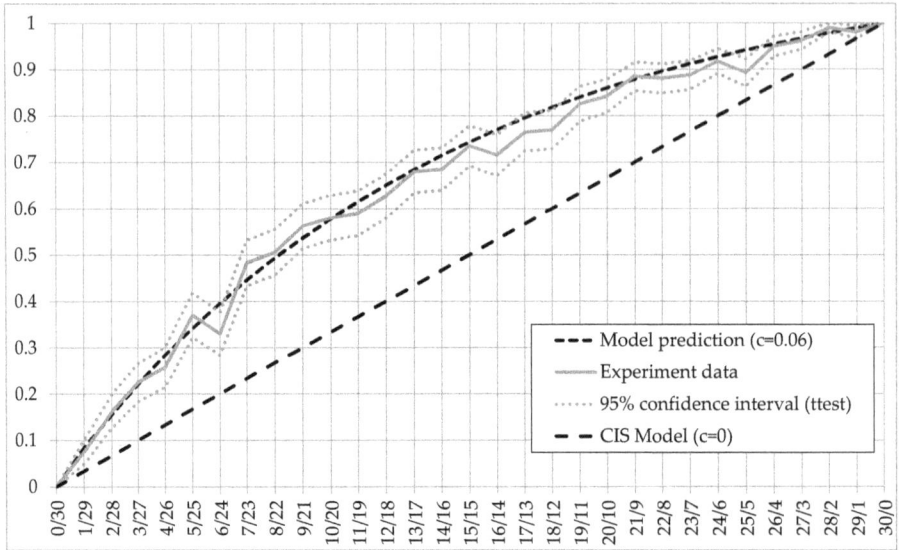

Fig. 6.12. Probabilities of converging to the superior stable state using decentralized selection. Tournament size has been set to $k = 4$, matrix C has not been predefined, but is given implicitly by environmental and selection properties; the according measured probability values are depicted in Fig. 6.11. As above, the average over the experimental results is given by the gray solid line.

selection operator. This is a clear sign that the confidence factor is not independent of the matrix C in this scenario. As a consequence, the chance for reaching the superior state is lower in this experiment than in the above experiments. This is a drawback for the model as it shows that the confidence factor cannot be estimated precisely in an artificial environment and simply transferred to a real scenario. Rather, the concrete configuration of C and c have to be figured out together in a single experimental setup. However, this observation is not unexpected as bridging the reality gap is a major problem in ER, and simple solutions are unlikely to be found. Nevertheless, the prediction model allows for a step-wise approximation from simulation to reality by gaining more and more information about C and c in every step as outlined in Fig. 6.1. It only has to be considered that the probability matrix C and the confidence factor c are dependent of each other and may not be measured separately.

6.4.3 Remarks on Evolution in the scope of the CIS Model

As shown above, the CIS model predicts for evolution (within the given framework) to not being able of improving a population toward an implicitly given goal in the long-term if no explicit fitness is considered by selection. However, scenarios using implicit fitness as a driving factor have been extensively and successfully studied in the field of open-ended artificial evolution [12]. Moreover, a recent EE study, closely related to the approaches considered in this thesis, has successfully utilized "implicit" evolution to evolve several behaviors in a swarm of robots [19]. These results do not pose a contradiction to the CIS model prediction; rather, the difference originates from the very strict notion of *implicity* used here.

For example, in [19] the authors consider an intrinsic energy level causing a robot to die or to be infertile if it does not find food in the environment to be "implicit" or "environment-driven". Using the above-defined prediction model, however, this would have to be considered a partially explicit selection. The CIS model allows for selection only to prefer individuals over others by properties built into the environment. A fertility-affecting energy level built into the environment, however, violates assumption 3 given in Sec. 6.1, as it has an impact on the robots' ability of controlling their chances of being selected by evolving their controllers. Such a mechanism would only be allowed in the CIS model if it evolved from scratch in an environment flexible enough to allow for this feature, but not when manually implemented from the beginning. Therefore, such results cannot be captured by the CIS model. The EIS model, on the other hand, can capture them by explicitly converting the energy level into a fitness value which the selection probability depends on. As argued above, by this strict definition even natural evolution includes non-implicit properties interfering with or being part of selection. A particularly strong property of this type is death which has not been evolved, and nevertheless is a major driving factor of natural evolution. (Note that this is a computer scientist's view on natural evolution, deliberately ignoring many biological details for the purpose of illuminating its artificial counterpart in ER.)

This shows that the CIS model is a rather theoretical concept, and its main usage is to build up the foundation of the EIS model. However, even when evolving in a strictly CIS-based scenario, improvements are not completely ruled out in the short or middle term. Undeniably, the probability of a chain of n "improvements" (i. e., convergences to a superior mutation) in a row decreases exponentially with n, and only one convergence to an inferior mutation is likely to cause a fall-back to a rudimentary behavior (as mutations are usually far more often harmful than beneficial). However, purely implicit selection pressure can keep the probability for harmful mutations to spread through the population sufficiently low to allow for evolution to improve for several convergence steps in a row. Therefore, fairly adapted behavior can be evolved before a fall-back to random behavior occurs. As an example, an experiment has been conducted using a setting similar to the approach described in Chap. 4, however, with the following changes made:

1. Reproduction is performed when two robots come spatially close to each other (reproduction distance is set to five times a robot's width). Without considering explicit fitness, one of the two implicitly selected robots is uniformly randomly selected to copy its controller to the other robot.

Fig. 6.13. Maze environment for the CIS scenario. The robots are supposed to find each other for mating, inducing an implicit selection factor; no explicit selection is performed.

2. After mating, the mating partners are relocated to random places uniformly distributed over all obstacle-free parts of the environment.
3. When a robot crashes into an obstacle, it is not turned, but just relocated by the same method as in Chap. 4.
4. The environment is shaped in a maze-like manner being more complex than the environments used so far, cf. Fig. 6.13.
5. The evolving population consists of 50 robots; evolution time is set to 600,000 simulation steps.

The purpose of changes (1) and (4) is to drive the selection process by implicitly allowing only robots to reproduce that have learned to traverse the maze. At least a driving forward behavior is required to trigger the reproduction process, but due to change (3), even driving forward does not lead to a good traversal strategy, but the environment clearly selects for a CA or Wall Following behavior. Change (2) prevents robots from reproducing many times with the same partner without having a need to find another one. The population size and the simulation time given in (5) have been established in preliminary tests. While explicit fitness is not considered during selection, it is still calculated to provide for a measure of "superiority" of the evolved behaviors. As "Maze Traversal" is expected to be selected for, again this measure is given by the explicit fitness defined in Sec. 6.4.1.

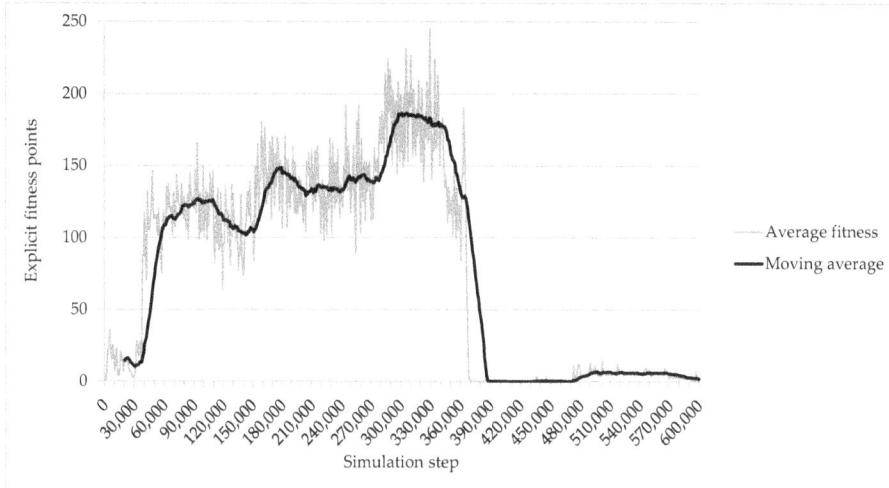

Fig. 6.14. Fitness progression in a CIS run involving fairly complex behavior. Evolution is capable of improving the behavior in several steps to a state of sophisticated Maze Traversal behavior (cf. Fig. 6.15) without any explicit selection factors. However, as predicted, selection pressure is to weak to prevent the evolved behavior from eventually being harmed by inferior mutations.

Running this scenario in simulation, virtually every time a driving behavior has temporarily been evolved. Typically, the evolved driving behaviors are very simple such as constantly driving-forward, but a small percentage of the runs has yielded more complex behaviors. Fig. 6.14 shows the fitness development of a very rare case where a sensitive behavior has been evolved capable of highly-effectively traversing the maze by turning to the left as well as to the right depending on the obstacles sensed. Fig. 6.15 shows one of the "best" (in terms of fitness) MARBs evolved in that run. The population is capable of several times improving the behavior raising average fitness to a maximum of nearly 250. However, as predicted, every run when simulated long enough at some point abruptly falls back to trivial behavior as the fitness chart shows exemplarily. Nevertheless, selection pressure in this scenario is still considerably higher than in a completely random search process where the occurrence of even a much simpler CA behavior would have been highly improbable, and its spread throughout the population virtually impossible.

Obviously, a completely implicit or purely environment-driven scenario cannot be used in a practical means to evolve behaviors in ESR. On the other hand, the above CIS scenario is capable of establishing a continuing selection pressure which shows the importance of implicit factors in evolutionary scenarios involving complex environments. Thus, the CIS model, which captures the most extreme case of environment-driven evolution, can provide valuable insights in the mechanisms active in more complex scenarios involving implicit and explicit factors.

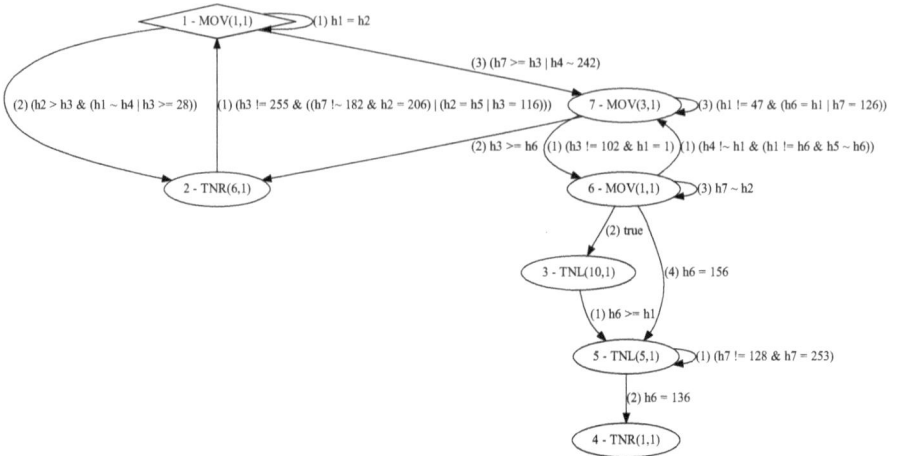

Fig. 6.15. One of the best MARBs evolved in a CIS run. The MARB has been simplified by omitting semantically irrelevant parts. It performs a sophisticated and effective Maze Traversal behavior allowing the robots to repeatedly reproduce. Still, the behavior involves more collisions than typical CA behaviors evolved by using an explicit CA fitness function. As collisions are not punished, they may well be part of an optimal behavior and the according accident simulations may even be exploited on purpose. (Cf. fitness progression in Fig. 6.14.)

6.5 Chapter Résumé

In this chapter, a mathematical model based on Markov chains has been introduced that can be utilized to estimate the probability that an ESR run (or any evolutionary run in a complex environment) will be successful, i. e., if selection will be capable of improving a population until a desired behavior is found. In contrast to existing work (cf. discussion in Chap. 2), the model considers both implicit fitness, which depends on potentially hidden environmental properties, and explicit fitness, which is calculated from environmental variables, and can be fuzzy, noisy or delayed. In complex environments both fitnesses may not reflect the closeness to a desired behavior perfectly. Furthermore, particularly implicit fitness is mostly given by the scenario and hard to influence. The model takes into account the chances for both the implicit and explicit part of selection to distinguish superior from inferior individuals. Based on this data, it calculates the probability that a population will eventually converge to a superior state, i. e., consisting of superior individuals only, in contrast to the undesired convergence to a state with only inferior individuals. Furthermore, the expected time to convergence in terms of the number of reproduction events required to reach a stable state can be calculated. The model is applicable to most common ESR scenarios including offline and centralized as well as online and decentralized approaches. Its purpose is to provide a deeper understanding of the mechanisms active during selection in ESR, and to help predicting the performance of ESR runs particularly in those cases where success is of critical importance or where failures can cause expensive damage. This is frequently, but not exclusively, the case in decentralized online, i. e., EE scenarios. Experiments in simulation show that the model predictions coincide with actual experimental data in the given EE context. There are no restrictions to controller types or evolutionary operators except for selection which is so far limited to the tournament-based reproduction scheme used throughout the thesis.

The model depends on a rather big number of input parameters which have a major influence on the results. As the model is only useful if these parameters are estimated correctly before an actual run (e. g., in simulation before performing a real-world run), future work is planned to further investigate how these parameters can be detected more accurately. To this end, one approach using a series of simulation runs has been presented in this chapter. However, it still has to be studied how well its results match the according outcome of an appropriate real-world scenario. Another open problem is the model's current inability to capture mutation directly. Rather, it is assumed that long phases without behavior-changing mutations exist, during which the model is applied. This assumption applies fairly well to many ESR scenarios (cf. discussion in Sec. 6.1), but not to all; and even for the scenarios it does apply to, it is not yet clear how different chains of mutations occurring in different runs change the model parameters and, thus, prediction quality. Future research is planned to address this problem bei either modeling mutation directly or finding another solution to correctly adapt the input parameters according to the occurring chains of mutations. Finally, tournament selection, while being a natural choice for ESR, is another constraint to the model that is not categorically required. In future, the model is planned to be extended to be applicable to selection methods different from tournament selection.

7 Conclusion

Preamble:

> *"First Law: A robot may not injure a human being or, through inaction, allow a human being to come to harm.*
> *Second Law: A robot must obey the orders given to it by human beings, except where such orders would conflict with the First Law.*
> *Third Law: A robot must protect its own existence as long as such protection does not conflict with the First or Second Law."*

The Three Laws of Robotics by Isaac Asimov [6]

Future developments in robotic hardware technology will soon yield low-cost autonomous robots for various kinds of private and professional applications. The control of such robots in every-day situations, and even more under critical circumstances such as those given by medical applications, will require a profound understanding of the complex interactions of robots with each other and with humans. This understanding firstly includes knowledge about the generation of algorithms in the first place that allow for complex robotic activities to be executed. In many cases this generation is highly complex even for a single robot, and gets harder when a swarm of robots is involved or when heterogeneous robots interact with each other. Secondly, dependable validation methods will be needed that go beyond mere algorithmic analysis and are capable of proving complex real-world properties of robot behaviors. For example, Isaac Asimov's famous *Three Laws of Robotics* are such complex properties that are undeniably desirable, but by today's techniques extremely hard to accomplish in a real robotic system, and hardly ever provable for all circumstances. The field of ER provides promising ideas that might bring forth methods to treat these issues in the future.

The contributions of this thesis to the field of ER can be arranged along the axes of (a) trustworthiness of both the behavior generation process and the resulting behaviors (mainly items 2., 3. and 4. in the chapter-wise listing below), and (b) generation of complex behaviors for autonomous robots (mainly item 3. below). Further contributions concern the field of ABS (item 1.). Following the chapter structure of the thesis, the major contributions can be summarized as follows:

1. Proposal, implementation and evaluation of a programming pattern for agent-based simulations intended to improve structuredness and code reusability in implementations by non-expert programmers (Chap. 3). Experiments with student test persons suggest that the proposed architecture implemented within the simulation program EAS yields more structured and reusable implementations than the state-of-the-art simulation programs MASON and NetLogo.

2. Proposal of a novel FSM-based control model for robots (Moore Automaton for Robot Behavior; MARB) and a corresponding decentralized online-evolutionary

framework, both applicable to various types of robots (Chap. 4). A comprehensive evaluation shows that robot behaviors of complexities comparable to the current state-of-the-art from literature can be evolved, by yielding fairly analyzable robot programs, thus, contributing to the trustworthiness of evolved behaviors in ER.

3. Proposal of a highly flexible genotype-phenotype mapping based on FSMs (Moore Automaton for Protected Translation; MAPT) which, by mimicking the process of DNA to protein translation in nature, can be evolved along with robot behavior, leading to an automatic adaptation of the evolutionary operators to a given search space structure (Chap. 5). It is shown by extensive evaluation that the approach can lead to the desired adaptation and furthermore to a significant improvement of the evolutionary outcomes. The question of the extent, to which the approach can support the evolution of truly complex behavior, has to be finally resolved in future work. Nevertheless, the flexibility of the genotype-phenotype mapping and the system's capability to control essential aspects of its own evolution in a highly recursive manner build a foundation for a broad range of applications and further research options (see below).

4. Proposal and evaluation of a formal framework for the prediction of success in evolutionary approaches involving complex environments (Chap. 6). By measuring the qualities of both fitness-based (explicit) and environmental (implicit) selection in variable states of abstractions from an intended real-world scenario (purely probabilistic estimations – simulation runs – laboratory experiments etc.), a probability of successful behavior evolution in the real environment can be estimated. Experiments with so far rather artificial simulation scenarios show that the model prediction matches the experimental outcome with a high precision.

In the following, a more detailed summary of the main topics of the thesis and an outlook to future work are given, according to the structure given by the two ER axes as well as the distinct ABS axis.

Agent-based simulation. There exists a large body of simulation frameworks, each of which provides different levels of complexity, various types of features and different matureness of documentation while requiring different degrees of user skills. This situation makes it difficult for non-expert users to choose an appropriate simulation framework and quickly begin programming individual simulations. The Simulation Plugin Interface (SPI) proposed in Chap. 3 is an architectural concept intended to provide programmers with a coherent interface which can be utilized to make existing simulations extendable in an inherently well-structured way. The SPI enforces users to program within these structures whatever type of simulation they are implementing, from simple extensions of existing simulations to complex simulation toolkits. In this way, errors and misconceptions can be found at an early stage, often at compilation time. Furthermore, by using the SPI, the software automatically gets reusable in a hierarchical way meaning that users can easily program extensions to simulations without having to alter any parts of the existing code. Additional implementations can be plugged in or out as desired and remain completely distinct from the core implementation. From this perspective, any implementation within the SPI architecture provides a new shell around the existing simulation core which, in turn, can be reused without being altered from "outside". Experiments with student test persons imply that implementations made by unskilled programmers, using the SPI as part of the presented ABS Framework EAS, are better-structured and more reusable than the according reference implementations using NetLogo and MASON. On the other hand, the time required by the test persons for implementing their simulation programs is comparable to the time required with MASON or NetLogo (for this statement, however, implementation parts have to be ignored which clearly rely on MASON's or NetLogo's far-developed auxiliary tools which have lacked a counterpart in the early EAS version used for the experiments).

To provide more specific insights in the advantages offered by the SPI architecture, more experiments with a larger group of test persons are required. There, particularly more comprehensive methods for measuring the implementation quality, as known from software engineering, are desirable to allow for a more detailed analysis and to possibly find further improvement possibilities. Furthermore, due to the universality of the SPI architecture, an implementation within most existing simulation programs is possible (Chap. 3 gives some clues to how this can be done). An integration in a simulation program such as MASON could yield insights about the accomplishments of the SPI architecture unbiased by the particular implementation given by EAS.

Trustworthiness. Most currently published successful approaches in ER are basing controllers on ANNs or similarly complex structures that, being Turing-complete in the most general case, provide for the greatest possible behavioral flexibility. However, Turing-completeness by itself implies that even from a solely algorithmic point of view arbitrary controllers cannot be automatically tested for any non-trivial properties. Moreover, even for the several constrained classes of ANNs commonly used in ER no analyzing methods for arbitrary controllers are known, neither automatic nor manual – using "common sense". Rather, behavioral properties are usually graded by observation and fitness measurement [86, 139], both of which methods provide statistic evidence only, and lack even a statement of confidence as it is unknown which untested situations might cause failures.

Chapters 4 and 5 show that much simpler controllers, namely MARB controllers which are completely representable by FSMs, can be used to evolve behaviors of comparable complexity as known from current literature (limited to completely evolutionary methods; cf. discussion in Chap. 2). For that purpose, an Embodied Evolution (EE) approach has been proposed and studied in simulation as well as with real robots. FSMs are both intuitively understandable for humans and provide a large body of theory including analyzing methods at an algorithmic level. Chap. 4 provides examples

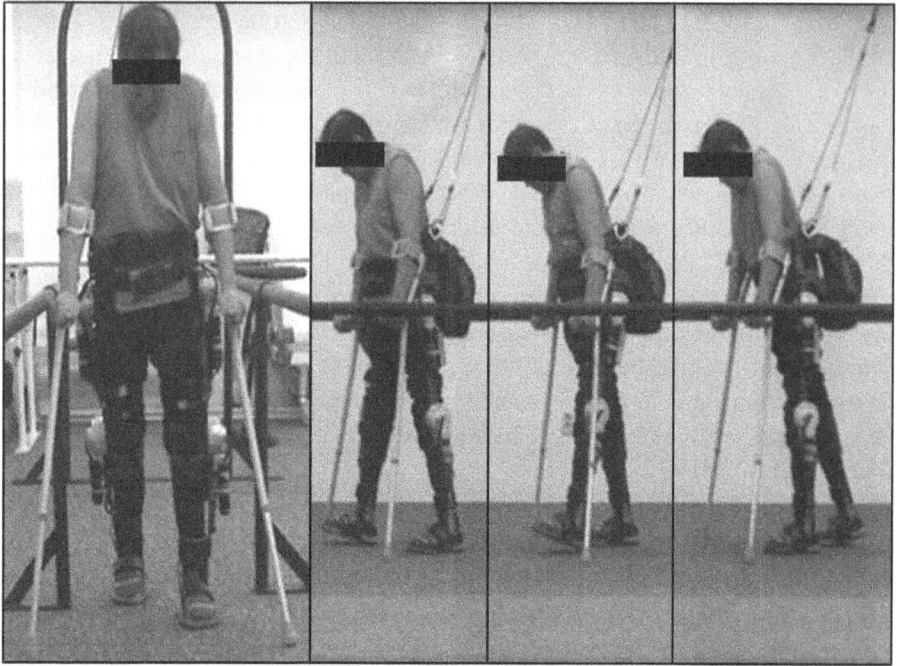

Fig. 7.1. MARB-controlled walk intention detection algorithm in a robotic exoskeleton with crutches. The figure shows a walking experiment with a paraplegic patient performed by a group of researchers from South Korea and the ETH Zürich.

of how basic behavioral properties can be proven by simply testing the reachability of states. Therefore, particularly for critical applications where failures may cause unforeseeable harm, MARBs can be used to generate trustworthy controllers. Recently, this has been done in a medical application by Jung et al. using MARBs to generate controllers for a robotic exoskeleton walking assistant with crutches for paraplegic patients [88], cf. Fig. 7.1.

However, the potentials of automatically analyzing MARBs have been discussed rather briefly in this thesis. Beyond mere state reachability more complex properties might be of interest such as "which sensor input sequences can lead to a certain state or operation". Such properties include the calculation of condition satisfiability which is in its essence equivalent to the *NP* complete propositional satisfiability (SAT) problem. Nevertheless, due to the small size and inherent structuredness of most conditions, this problem is expected to be treatable fairly easily, for example using SAT solvers which are, on average, highly efficient today. Classic FSM techniques such as minimization of automata can also be considered when analyzing MARBs. Future work should involve studying these issues to enhance the trustworthiness of the model. However, looking one step further, these aspects of trustworthiness still cover the algorithmic level only. The equally important question of how an algorithmic analysis can be transferred into real-world behavioral assertions remains, to a great extent, an open issue and another major subject for future research.

A different aspect of trustworthiness is addressed in Chap. 6, where the question is treated, how successful an evolutionary run will be when performed in a real-world environment. As real-world experiments can be very expensive, providing an accurate success prediction before the actual run can be of great importance. There has been a lot of theoretical work aiming at the calculation of the expected success of an evolutionary run in the fields of classic EC, spatially structured evolutionary algorithms, evolutionary optimization under uncertainty or in noisy environments and ER; on the other hand, great effort has been made by evolutionary biologists to understand population dynamics in nature (cf. discussion in Sec. 2.1.9 of Chap. 2). However, to the knowledge of the author, there exists so far no theoretical model including both implicit (environmental) and explicit (fitness-based in terms of classic EC) selection in the calculation of success. The combination of these two types of selection is considered typical to ESR and occurs neither in classic EC nor in evolutionary biology. The proposed model provides a prediction method based on Markov chains driven by data from previous simulation or laboratory runs. It bridges the reality gap by abstraction from scenario details and focussing on variables expected to be of major importance to the success of a run. Results show that the model is well applicable to several example scenarios.

However, so far experiments have been performed solely in simple simulated environments and by using selection methods designed to closely suit the model requirements. It remains to be shown that the model is applicable to more realistic real-robot scenarios, too. Furthermore, mutation is not yet explicitly considered by the model meaning that the prediction is constrained to periods without mutation. As different chains of mutation can lead to greatly differing results in terms of behavioral quality, including mutation into the model is another major concern for future work.

Complex behavior generation. One of the great open problems in ER is the question of weather or not it is possible to evolve truly complex behavior. More precisely, for behaviors exceeding a certain complexity, no purely evolutionary experiment so far has been capable to manage the bootstrap from an initial population of robots with trivial or random controllers to a population of robots performing the desired behavior. On the other hand, natural evolution has been capable of evolving organisms of much higher complexity than this threshold which suggests that the natural mechanisms might not be sufficiently well exploited in ER.

In Chap. 5, an evolutionary setting is proposed that evolves the genotype-phenotype mapping (GPM) along with the behavior, replicating nature's mechanism of evolving the translation process from DNA to proteins along with the evolution of the organisms' phenotypes. Changing the GPM implies a change of the effects of genotypic mutations to the resulting robot behaviors on the phenotype level. Furthermore, as the

GPM is encoded among the behavioral properties in a robot's genotype, it is translated together with them by "an earlier version of itself" to become part of a robot's phenotype (which consists of the behavior defined by a MARB and the GPM defined by a MAPT). Therefore, adapting the effects of evolutionary operations on the phenotypic level as described above recursively includes adapting the effects of these operations on the GPM. In this way, GPMs and, in consequence, mutation strategies for both the evolution of behaviors and of the GPM can be coevolved to match the search space features given by the target behavior to evolve and the according environmental properties.

While there exist approaches to adapt the effects of evolutionary operations during evolution (the most prominent being the Evolution Strategies, cf. discussions in Chapters 2 and 5), they usually require fixed top-level operations, i. e., "the meta-mutation operation which changes the effects of the mutation operation remains unchanged". Using the recursively self-adapting GPM structure proposed here, there is only one level of evolutionary operations rather than a hierarchy of meta-operations, thus allowing for an adaptation of the effects of these operations in a highly flexible way. Experiments in simulated EE show that the approach allows for the GPM to successfully adapt to given search spaces as intended, thus increasing evolvability. Furthermore, for the behaviors evolved in Chap. 4 using classic static evolution, a significantly higher success rate has been achieved with the evolvable GPM. Therefore, the approach using an evolvable GPM provides a robust foundation for evolving these behaviors in an EE setting.

Nevertheless, for several tested behaviors of higher complexities, for example, orientation in a dynamically changing environment, so far no successful results have been achieved. This suggests that the observed improvements might originate not only from the desired increase in evolvability due to the evolvable GPM, but primarily from effects such as an increased population diversity. One reason for a potentially insufficient exploitation of the GPM flexibility might be the syntactical and computational constraints induced by the usage of FSMs. A good starting point for further research is, therefore, to transfer the approach to ANN controllers. As these have fewer syntactical restrictions and a greater computational flexibility than FSMs they might allow for a smoother evolution of the GPM (Chap. 5 discusses some important issues concerning the implementation of the approach using ANNs). For this purpose the above-mentioned structural advantages of FSMs would have to be given up, but hopefully insights from such studies might eventually be transferred back to FSM-based evolution.

Furthermore, the approach has shown to improve evolvability in a specifically designed test scenario. Therefore, another way of improving the adaptation of GPMs can be to more accurately investigate the test scenario and carefully transfer it into a more realistic scenario.

Finally, the EE approach followed in this thesis might be unfavorable for a proper functioning of evolvable GPMs as it implies constraints to selection and fitness calculation. A centralized and offline evolutionary approach might provide means to more accurately select for desired behavioral and translator properties. In principle, the proposed evolvable GPM is not even limited to ER, but suitable for many evolutionary settings. Overall, the approach is promising to enhance evolutionary systems in their ability to generate complexity, however, solving the problem of adjusting it properly might still require a lot of effort.

Closing statement

The field of ER and its subfield ESR offer great potentials for the automatic creation of robotic behaviors. However, the road which has to be followed until evolution of reliable complex autonomous robotic systems may get into reach still involves obstacles and unresolved problems. While it is likely that significant progress will be made in near future, it is unforeseeable from today's perspective if systems comparable to nature's simplest animal "robots" such as flies or spiders will ever be achievable by artificial processes. As long as the natural example seems unattainable, a noticeable thought is that not killing a fly or a spider may always be a greater contribution to complexity in the world than putting the highest possible effort in building a sophisticated robot. Considering this and setting it into relation to the actual amount of animals, including humans, coming to harm from consequences of human progress, elucidates the high moral responsibility which technological research overall and ER in particular have. Apart from mere progress, scientists should be aware of the implications of their findings and responsibly put them into perspective before letting them enter a self-organized and market-driven application phase. If used wisely, major advancements in every-day life as well as in many important fields of research and application can be expected through ER in the years and decades to come.

References

[1]   A. Acerbi, D. Marocco, and S. Nolfi. Social facilitation on the development of foraging be-
        haviors in a population of autonomous robots. In *Advances in Artificial Life*, volume 4648 of
        Lecture Notes in Computer Science, pages 625–634. Springer, 2007.

[2]   C. Ampatzis, E. Tuci, V. Trianni, and M. Dorigo. Evolving communicating agents that integrate
        information over time: a real robot experiment. In *CD–ROM Proceedings of the Seventh Inter-
        national Conference on Artificial Evolution*. Springer, 2005.

[3]   C. Ampatzis, E. Tuci, V. Trianni, and M. Dorigo. Evolution of signalling in a group of robots
        controlled by dynamic neural networks. In *Proceedings of the second workshop on swarm
        robotics*. Springer, 2006.

[4]   C. Ampatzis, E. Tuci, V. Trianni, and M. Dorigo. Evolution of signaling in a multi-robot system:
        categorization and communication. In *Evolution of communication and language in embodied
        agents*, pages 161–178. Springer, 2010.

[5]   D. V. Arnold. *Evolution strategies in noisy environments – a survey of existing work*, pages
        239–250. Theoretical aspects of evolutionary computing. Springer, 2001.

[6]   I. Asimov. *Runaround*. Street & Smith, 1942.

[7]   W. Banzhaf, P. Nordin, and M. Olmer. Generating adaptive behavior using function regression
        within genetic programming and a real robot. In *2nd international conference on genetic
        programming*, pages 35–43. Morgan Kaufmann, 1997.

[8]   J. Barcelo, E. Codina, J. Casas, J. L. Ferrer, and D. Garcia. Microscopic traffic simulation: a tool
        for the design, analysis and evaluation of intelligent transport systems. *Journal of Intelligent
        and Robotic Systems*, 41:173–203, 2005.

[9]   G. Barlow, C. Oh, and E. Grant. Incremental evolution of autonomous controllers for un-
        manned aerial vehicles using multi-objective genetic programming. In *IEEE conference on
        cybernetics and intelligent systems*, pages 689 – 694, 2004.

[10]   G. J. Barlow, L. S. Mattos, and E. Grant. Transference of evolved unmanned aerial vehicle
        controllers to a wheeled mobile robot. In *Proceedings of the IEEE international conference on
        robotics*, pages 704–720. MIT Press, 2004.

[11]   J. Bay. *The thought works anthology: essays on software technology and innovation*, chapter
        6 – Object calisthenics. O'Reilly Series. Pragmatic Bookshelf, 2008.

[12]   M. A. Bedau, J. S. McCaskill, N. H. Packard, S. Rasmussen, C. Adami, D. G. Green, T. Ikegami,
        K. Kaneko, and T. S. Ray. Open problems in artificial life. *Artificial Life*, 6:363–376, 2000.

[13]   H.-G. Beyer and H.-P. Schwefel. Evolution strategies: a comprehensive introduction. *Natural
        Computing*, 1:3–52, 2002.

[14]   E. Bonabeau, G. Theraulaz, and M. Dorigo. *From natural to artificial systems*. Oxford Univer-
        sity Press, 1999.

[15]   M. Bonn. *JoSchKa: jobverteilung in heterogenen und unzuverlässigen Umgebungen*. PhD
        thesis, Universität Karlsruhe (TH), 2008.

[16]   M. Bonn and H. Schmeck. The joschka system: organic job distribution in heterogeneous and unreliable environments. In *Proceedings of the 23rd international conference on architecture of computing systems*, pages 73–86. Springer, 2010.

[17]   V. Braitenberg. *Vehicles: experiments in synthetic psychology*. MIT Press, 1984.

[18]   J. Branke and H. Schmeck. Evolutionary design of emergent behavior. In *Organic computing*, Understanding Complex Systems, 2008.

[19]   N. Bredeche and J.-M. Montanier. Environment-driven embodied evolution in a population of autonomous agents. In *Parallel problem solving from nature*, pages 290–299. Springer, 2010.

[20]   G. Buason, N. Bergfeldt, and T. Ziemke. Brains, bodies, and beyond: competitive co-evolution of robot controllers, morphologies and co-evolution. *Genetic Programming and Evolvable Machines*, 6:25–51, 2005.

[21]   S. Burbeck. Applications programming in smalltalk-80: How to use model-view-controller (mvc), 1987.

[22]   J. Casas, J. L. Ferrer, D. Garcia, J. Perarnau, and A. Torday. Traffic simulation with aimsun. In *Fundamentals of Traffic Simulation*, volume 145 of *International Series in Operations Research and Management Science*, pages 173–232. Springer, 2010.

[23]   C. Castle and A. Crooks. Principles and concepts of Agent-Based modelling for developing geospatial simulations. *UCL working papers series*, 110, 2006.

[24]   W. K. Chan, Y.-J. Son, and C. M. Macal. Agent-based simulation tutorial – simulation of emergent behavior and differences between agent-based simulation and discrete-event simulation. In *Proceedings of the 2010 Winter Simulation Conference*, pages 135–150, 2010.

[25]   A. L. Christensen, R. O'Grady, and M. Dorigo. Parallel task execution, morphology control and scalability in a swarm of self-assembling robots. In *Proceedings of the 9th Conference on Autonomous Robots and Competitions*, 2009.

[26]   D. Cliff and G. Miller. Tracking the red queen: Measurements of adaptive progress in co-evolutionary simulations. In *Proceedings of the Third European Conference on Artificial Life: Advances in Artificial Life*, pages 200–218, 1995.

[27]   D. Cliff and G. F. Miller. Co-evolution of pursuit and evasion ii: Simulation methods and results. In *From Animals to Animats 4: Proceedings of the Fourth International Conference on Simulation of Adaptive Behavior*, pages 506–515, 1996.

[28]   D. Cliff, I. Harvey, and P. Husbands. Explorations in evolutionary robotics. *Adaptive Behavior*, 2, 1993.

[29]   N. Collier, T. Howe, and M. North. Onward and upward: the transition to repast 2.0. In *First Annual North American Association for Computational Social and Organizational Science Conference*, 2003.

[30]   C. Constantinescu, S. Kornienko, O. Kornienko, and U. Heinkel. An agent-based approach to support the scalability of change propagation. In *Proceedings of the 17th International Conference on Parallel and Distributed Computing Systems*, 2004.

[31] M. Daniels. Integrating simulation technologies with swarm. In *Proceedings of the Workshop on Agent Simulation: Applications, Models, and Tools*, 1999.

[32] C. Darwin. *On the Origin of Species – A Facsimile of the First Edition*. Harward University Press, 1859.

[33] R. Dawkins. The evolution of evolvability. In *Artificial Life Proceedings*, 1987.

[34] R. Dawkins. *The selfish gene*. Oxford university press, 2006.

[35] K. De Jong. Evolutionary computation: a unified approach. In *Proceedings of the 2008 conference on genetic and evolutionary computation*, pages 2245–2258. ACM, 2008.

[36] E. do Valle Simoes and D. A. C. Barone. Predation: An approach to improving the evolution of real robots with a distributed evolutionary controller. In *IEEE International Conference on Robotics and Automation*, pages 664–669, 2002.

[37] M. Dorigo, V. Trianni, E. Sahin, R. Gross, T. H. Labella, G. Baldassarre, S. Nolfi, J.-L.Deneubourg, F.Mondada, D. Floreano, and L. M. Gambardella. Evolving self-organizing behaviors for a swarm-bot. *Autonomous Robots*, 17:223–245, 2004.

[38] M. Duarte, S. Oliveira, and A. Christensen. Hierarchical evolution of robotic controllers for complex tasks. In *IEEE International Conference on Development and Learning and Epigenetic Robotics*, pages 1–6, 2012.

[39] D. J. Earl and M. W. Deem. Evolvability is a selectable trait. *Proceedings of the National Academy of Sciences of the United States of America*, 101:11531–11536, 2004.

[40] M. Ebner and A. Zell. Evolving a behavior-based control architecture-From simulations to the real world. In *Proceedings of the Genetic and Evolutionary Computation Conference*, volume 2, 1999.

[41] Eclipse-Foundation, 2014. http://www.eclipse.org, retrieved on November 11th 2014.

[42] P. Eggenberger. Evolving morphologies of simulated 3d organisms based on differential gene expression. In *Proceedings of the fourth european conference on Artificial Life*, pages 205–213. MIT Press, 1997.

[43] V. Feldman. Evolvability from learning algorithms. In *40th annual ACM symposium on theory of computing*, 2008.

[44] D. Filliat, J. Kodjabachian, and J. a. Meyer. Incremental evolution of neural controllers for navigation in a 6-legged robot. In *Proceedings of the fourth International Symposium on Artificial Life and Robotics*. Univ. Press, 1999.

[45] D. Floreano and F. Mondada. Evolution of homing navigation in a real mobile robot. *IEEE Transactions on Systems, Man, and Cybernetics–Part B: Cybernetics*, 1996.

[46] D. Floreano, P. Dürr, and C. Mattiussi. Neuroevolution: from architectures to learning. *Evolutionary Intelligence*, 1:47–62, 2008.

[47] D. Floreano, P. Husbands, and S. Nolfi. *Evolutionary Robotics in Springer Handbook of Robotics*. Springer, 2008.

[48]   D. Fogel. *Evolutionary computation: toward a new philosophy of machine intelligence.* IEEE series on computational intelligence. Wiley, 2006.

[49]   L. J. Fogel. *Intelligence through simulated evolution: forty years of evolutionary programming.* Wiley, 1999.

[50]   L. J. Fogel, A. J. Owens, and M. J. Walsh. *Artificial Intelligence through Simulated Evolution.* Wiley, 1966.

[51]   L. J. Fogel, P. J. Angeline, and D. B. Fogel. An evolutionary programming approach to self-adaptation on finite state machines. In *Proceedings of the Fourth Annual Conference on Evolutionary Programming*, pages 355–365. MIT Press, 1995.

[52]   D. J. Futuyma. *Evolutionary Biology.* Sinauer Associates Inc., 1998.

[53]   O. Gigliotta and S. Nolfi. Formation of spatial representations in evolving autonomous robots. In *IEEE Symposium on Artificial Life*, pages 171–178, 2007.

[54]   D. E. Goldberg and K. Deb. A comparative analysis of selection schemes used in genetic algorithms. In *Foundations of Genetic Algorithms*, pages 69–93. Morgan Kaufmann, 1991.

[55]   D. E. Goldberg and J. Richardson. Genetic algorithms with sharing for multimodal function optimization. In *Proceedings of the second international conference on genetic algorithms and their application*, pages 41–49. L. Erlbaum Associates Inc., 1987.

[56]   F. Gomez and R. Miikkulainen. Incremental evolution of complex general behavior. *Adaptive Behavior*, 5:317–342, 1997.

[57]   S. Goss, S. Aron, J. L. Deneubourg, and J. M. Pasteels. Self-organized shortcuts in the argentine ant. *Naturwissenschaften*, 76:579–581, 1989.

[58]   C. Grinstead and L. Snell. *Introduction to Probability.* American Mathematical Society, 1997.

[59]   R. Groß and M. Dorigo. Evolution of solitary and group transport behaviors for autonomous robots capable of self-assembling. *Adaptive Behavior*, 16:285–305, 2008.

[60]   R. Groß and M. Dorigo. Cooperative transport of objects of different shapes and sizes. In *ANTS 2004*, Lecture Notes in Computer Science. Springer, 2004.

[61]   R. Gross and M. Dorigo. Towards group transport by swarms of robots. *International Journal of Bio-Inspired Computing*, 1:1–13, 2009.

[62]   G. Harik, E. Cantu-Paz, D. E. Goldberg, and B. L. Miller. The gambler's ruin problem, genetic algorithms, and the sizing of populations. *Evolutionary Computation*, 7:231–253, 1999.

[63]   C. Hartland and N. Bredeche. Evolutionary robotics, anticipation and the reality gap. In *IEEE International Conference on Robotics and Biomimetics*, pages 1640–1645. IEEE Computer Society, 2006.

[64]   I. Harvey, P. Husbands, and D. Cliff. Seeing the light: Artificial evolution, real vision. In *From Animals to Animates 3: Proceedings of the 3rd International Conference on Simulation of Adaptive Behavior*, pages 392–401, 1994.

[65]   M. Heindl. *Automatenbasierte evolution in realen roboterschwärmen (diploma thesis).* Karlsruhe Institute of Technology, 2011.

[66]   M. Heindl, L. König, and H. Schmeck. FSM-based evolution in a swarm of real wanda robots. Techreport, Karlsruhe Institute of Technology (AIFB), 2012.

[67]   R. Helaoui. *Morphological Development of Artificial Embodied Organisms under the Control of Gene Regulatory Networks (diploma thesis)*. Karlsruhe Institute of Technology, 2008.

[68]   S. Hettiarachchi. *Distributed Evolution for Swarm Robotics (PhD thesis)*. University of Wyoming, 2007.

[69]   S. Hettiarachchi. An evolutionary approach to swarm adaptation in dense environments. In *International Conference on Control Automation and Systems*, pages 962–966. IEEE, 2010.

[70]   S. Hettiarachchi and W. Spears. DAEDALUS for agents with obstructed perception. In *IEEE Mountain Workshop on Adaptive and Learning Systems*. Springer, 2006.

[71]   G. Hinton and S. Nowlan. How learning can guide evolution. *Complex Systems*, 1:495–502, 1987.

[72]   R. Hirschorn. Invertibility of nonlinear control systems. *SIAM Journal on Control and Optimization*, 17:289–297, 1979.

[73]   R. Hirschorn. Invertibility of multivariable nonlinear control systems. *IEEE Transactions on Automatic Control*, 24:855–865, 1979.

[74]   F. Hoffmann and J. C. S. Z. Montealegre. Evolution of a tactile wall-following behavior in real time. In *The 6th Online World Conference on Soft Computing in Industrial Applications*, pages 10–24, 2001.

[75]   F. Hoffmann and G. Pfister. Evolutionary learning of a fuzzy control rule base for an autonomous vehicle. In *Proceedings of the Fifth International Conference on Information Processing and Management of Uncertainty in Knowledge-Based Systems*, pages 659–664, 1996.

[76]   D. R. Hofstadter. *Gödel, Escher, Bach: An Eternal Golden Braid*. Basic Books, 1979.

[77]   J. Holland. *Adaptation in Natural and Artificial Systems*. University of Michigan Press, 1975.

[78]   J. E. Hopcroft, R. Motwani, and J. D. Ullman. *Introduction to Automata Theory, Languages, and Computation*. Addison-Wesley, 2006.

[79]   R. Horak. *Telecommunications and data communications handbook*. Wiley, 2007.

[80]   G. Hornby, S. Takamura, O. Hanagata, M. Fujita, and J. Pollack. Evolution of controllers from a high-level simulator to a high dof robot. In *Proceedings of the Third International Conference on Evolvable Systems: From Biology to Hardware*, pages 80–89, 2000.

[81]   G. Hornby, A. Globus, D. Linden, and J. Lohn. Automated antenna design with evolutionary algorithms. In *Proceedings of the 2006 AIAA Space Conference*, 2006.

[82]   H. Hyötyniemi. Turing machines are recurrent neural networks. In *Proceedings of STeP'96*, pages 13–24. Finnish Artificial Intelligence Society, 1996.

[83]   N. Jakobi. Evolutionary robotics and the radical envelope-of-noise hypothesis. *Adaptive Behavior*, 6:325–368, 1997.

[84]   N. Jakobi and M. Quinn. Some problems (and a few solutions) for open-ended evolutionary robotics. In *Evolutionary Robotics*, volume 1468 of *Lecture Notes in Computer Science*, pages 108–122. Springer, 1998.

[85]   N. Jakobi, P. Husbands, and I. Harvey. Noise and the reality gap: The use of simulation in evolutionary robotics. In *Advances in artificial life: proceedings of the third european conference on artificial life*, pages 704–720. Springer, 1995.

[86]   Y. Jin. A comprehensive survey of fitness approximation in evolutionary computation. *Soft Computing - A Fusion of Foundations, Methodologies and Applications*, 9:3–12, 2005.

[87]   Y. Jin, M. Olhofer, and B. Sendhoff. Managing approximate models in evolutionary aerodynamic design optimization. In *In Proceedings of the IEEE Congress on Evolutionary Computation*, pages 592–599. IEEE, 2001.

[88]   J. Jung, I. Jang, R. Riener, and H. Park. Walking intent detection algorithm for paraplegic patients using a robotic exoskeleton walking assistant with crutches. *International Journal of Control, Automation, and Systems*, 10:954–962, 2012.

[89]   M. Kaiser, H. Friedrich, R. Buckingham, K. Khodabandehloo, and S. Tomlinson. Towards a general measure of skill for learning robots. In *Proceedings of the 5th European Workshop on Learning Robots*, 1996.

[90]   C. Kambhampati, R. Craddock, M. Tham, and K. Warwick. Inverting recurrent neural networks for internal model control of nonlinear systems. In *Proceedings of the American Control Conference*, volume 2, pages 975–979, 1998.

[91]   S. Kamio and H. Iba. Adaptation technique for integrating genetic programming and reinforcement learning for real robots. *IEEE Transactions on Evolutionary Computation*, 9:318 – 333, 2005.

[92]   R. Keller and W. Banzhaf. Genetic programming using genotype-phenotype mapping from linear genomes into linear phenotypes. In *Proceedings of the First Annual Conference on Genetic Programming*, pages 116–122. MIT Press, 1996.

[93]   S. Kernbach, E. Meister, F. Schlachter, K. Jebens, M. Szymanski, J. Liedke, D. Laneri, L. Winkler, T. Schmickl, R. Thenius, P. Corradi, and L. Ricotti. Symbiotic robot organisms: REPLICATOR and SYMBRION projects. In *Proceedings of the 8th Workshop on Performance Metrics for Intelligent Systems*, pages 62–69. ACM, 2008.

[94]   S. Kernbach, E. Meister, O. Scholz, R. Humza, J. Liedke, L. Ricotti, J. Jemai, J. Havlik, and W. Liu. Evolutionary robotics: The next-generation-platform for on-line and on-board artificial evolution. In *IEEE Congress on Evolutionary Computation*, 2009.

[95]   D. Kessler, H. Levine, D. Ridgway, and L. Tsimring. Evolution on a smooth landscape. *Journal of Statistical Physics*, 87:519–543, 1997.

[96]   A. Kettler, M. Szymanski, J. Liedke, and H. Wörn. Introducing wanda – a new robot for research, education, and arts. In *International Conference on Intelligent Robots and Systems*, 2010.

[97]   D. Keymeulen, M. Iwata, Y. Kuniyoshi, and T. Higuchi. Online evolution for a self-adapting robotic navigation system using evolvable hardware. *Artificial Life*, 4:359–393, 1998.

[98]   M. Kimura. *The Neutral Theory of Molecular Evolution*. Cambridge University Press, 1985.

[99]   J. Kodjabachian and J.-A. Meyer. Evolution and development of neural controllers for locomo-
       tion, gradient-following, and obstacle-avoidance in artificial insects. *Neural Networks, IEEE
       Transactions on*, 9:796–812, 1998.

[100] L. König. *A model for developing behavioral patterns on multi-robot organisms using con-
       cepts of natural evolution (diploma thesis)*. University of Stuttgart, 2007.

[101] L. König and D. Pathmaperuma. EAS project page at sourceforge, 2012. http://sourceforge.
       net/projects/eas-framework, retrieved on November 11th 2014.

[102] L. König and H. Schmeck. Evolving collision avoidance on autonomous robots. In *Biologically
       Inspired Collaborative Computing*, pages 85–94, 2008.

[103] L. König and H. Schmeck. A completely evolvable genotype-phenotype mapping for evolu-
       tionary robotics. In *International Conference on Self-Adaptive and Self-Organizing Systems*,
       pages 175–185, 2009.

[104] L. König and H. Schmeck. Evolvability in evolutionary robotics: Evolving the genotype-
       phenotype mapping. In *Proceedings of the Fourth IEEE International Conference on Self-
       Adaptive and Self-Organizing Systems*, pages 259–260. IEEE, 2010.

[105] L. König, K. Jebens, S. Kernbach, and P. Levi. Stability of online and onboard evolving of
       adaptive collective behavior. In *European Robotics Symposium*, Springer Tracts in Advanced
       Robotics, 2008.

[106] L. König, S. Mostaghim, and H. Schmeck. Online and onboard evolution of robotic behaiv-
       ior using finite state machines. In *Proceedings of the 8th International Conference on Au-
       tonomous Agents and Multiagent Systems*, volume 2, 2009.

[107] L. König, S. Mostaghim, and H. Schmeck. Decentralized evolution of robotic behavior using
       finite state machines. *International Journal of Intelligent Computing and Cybernetics*, 2:695–
       723, 2009.

[108] L. König, S. Mostaghim, and H. Schmeck. A Markov-chain-based model for success pre-
       diction of evolution in complex environments. In *Proceedings of the 3rd International Joint
       Conference on Computational Intelligence*, pages 90–102. IEEE, 2011.

[109] L. König, D. Pathmaperuma, F. Vogel, and H. Schmeck. Introducing the simulation plugin
       interface and the eas framework with comparison to two state-of-the-art agent simulation
       frameworks. In *Proceedings of the 2012 Winter Simulation Conference*, pages 412:1–412:13.
       Winter Simulation Conference, 2012.

[110] S. Kornienko, O. Kornienko, and P. Levi. About nature of emergent behavior in micro-systems.
       In *Proceedings of the International Conference on Informatics in Control, Automation and
       Robotics*, pages 33–40, 2004.

[111] J. R. Koza. *Genetic Programming – On the Programming of Computers by Means of Natural
       Selection*. MIT Press, 1992.

[112] G. E. Krasner and S. T. Pope. A cookbook for using the model-view controller user interface
       paradigm in smalltalk-80. *Journal of Object-Oriented Programming*, 1:26–49, 1988.

[113] C. R. Kube and H. Zhang. Collective robotics: From social insects to robots. *Adaptive Behavior*, pages 189–218, 1993.

[114] J. Lehman and K. O. Stanley. Evolvability is inevitable: Increasing evolvability without the pressure to adapt. *PLoS ONE*, 8, 2013.

[115] H. Lipson and J. Pollack. Automatic design and manufacture of robotic lifeforms. *Nature*, 406: 974–978, 2000.

[116] T. Lochmatter, P. Roduit, C. Cianci, N. Correll, J. Jacot, and A. Martinoli. SwisTrack – A Flexible Open Source Tracking Software for Multi-Agent Systems, 2008.

[117] S. Luke, C. Cioffi-Revilla, L. Panait, K. Sullivan, and G. Balan. Mason: A multiagent simulation environment. *Simulation*, 81:517–527, 2005.

[118] H. H. Lund and O. Miglino. From simulated to real robots. In *Proceedings of the IEEE International Conference on Evolutionary Computation*, pages 362–365. IEEE, 1996.

[119] S. Lytinen and S. Railsback. The evolution of agent-based simulation platforms: A review of NetLogo 5.0 and ReLogo. In *Proceedings of the Fourth International Symposium on Agent-Based Modeling and Simulation*, 2012.

[120] C. M. Macal and M. J. North. Tutorial on agent-based modelling and simulation. *Journal of Simulation*, 4:151–162, 2010.

[121] D. Marocco and D. Floreano. Active vision and feature selection in evolutionary behavioral systems. In *From Animals to Animats*, Lecture Notes in Computer Science, pages 247–255. Springer, 2002.

[122] D. Marocco and S. Nolfi. Origins of communication in evolving robots. In *From Animals to Animats 9*, Lecture Notes in Computer Science, pages 789–803. Springer, 2006.

[123] M. Matarić and D. Cliff. Challenges in evolving controllers for physical robots. *Robotics and Autonomous Systems*, 19:67 – 83, 1996.

[124] V. Matellán, C. Fernández, and J. M. Molina. Genetic learning of fuzzy reactive controllers. *Robotics and Autonomous Systems*, 25:33–41, 1998.

[125] H. Maturana and F. Varela. *The Tree of Knowledge: The Biological Roots of Human Understanding*. Shambhala, 1987.

[126] S. Merkel, L. König, and H. Schmeck. Age based controller stabilization in evolutionary robotics. In *2nd World Congress on Nature and Biologically Inspired Computing*, pages 84–91. IEEE, 2010.

[127] J.-A. Meyer, P. Husbands, and I. Harvey. Evolutionary robotics: A survey of applications and problems. In *Evolutionary Robotics*, pages 1–21. Springer, 1998.

[128] O. Miglino, D. Denaro, G. Tascini, and D. Parisi. Detour behavior in evolving robots: Are internal representations necessary? In *Proceedings of the First European Workshop on Evolutionary Robotics*, pages 59–70. Springer, 1998.

[129] N. Minar. *The swarm simulation system : a toolkit for building multi-agent simulations*. Santa Fe Institute, 1996.

[130] J.-M. Montanier and N. Bredeche. Embedded evolutionary robotics: The (1+ 1)-restart-online adaptation algorithm. In *New Horizons in Evolutionary Robotics*, pages 155–169. Springer, 2011.

[131] D. C. Montgomery. *Design and Analysis of Experiments*. Wiley, 2008.

[132] J. M. Moore and P. K. McKinley. Evolution of an amphibious robot with passive joints. In *IEEE Congress on Evolutionary Computation*, pages 1443–1450. IEEE, 2013.

[133] J.-B. Mouret and S. Doncieux. Incremental evolution of animats: Behaviors as a multi-objective optimization. In *From Animals to Animats 10*, volume 5040 of *Lecture Notes in Computer Science*, pages 210–219. Springer, 2008.

[134] J.-B. Mouret and S. Doncieux. Overcoming the bootstrap problem in evolutionary robotics using behavioral diversity. In *IEEE Congress on Evolutionary Computation*, pages 1161–1168, 2009.

[135] H. Nakamura, A. Ishiguro, and Y. Uchilkawa. Evolutionary construction of behavior arbitration mechanisms based on dynamically-rearranging neural networks. In *Congress on Evolutionary Computation*, volume 1, pages 158 –165, 2000.

[136] U. Nehmzow. Physically embedded genetic algorithm learning in Multi-Robot scenarios: The PEGA algorithm, 2002.

[137] A. Nelson, E. Grant, J. Galeotti, and S. Rhody. Maze exploration behaviors using an integrated evolutionary robotics environment. *Robotics and Autonomous Systems*, 46:159 – 173, 2004.

[138] A. L. Nelson and E. Grant. Using direct competition to select for competent controllers in evolutionary robotics. *Robotics and Autonomous Systems*, 54:840 – 857, 2006.

[139] A. L. Nelson, G. J. Barlow, and L. Doitsidis. Fitness functions in evolutionary robotics: a survey and analysis. *Robotics and Autonomous Systems*, 57:345–370, 2009.

[140] C. Nikolai and G. Madey. Tools of the trade: A survey of various agent based modeling platforms, 2009. http://jasss.soc.surrey.ac.uk/12/2/2.html, retrieved on November 11th 2014.

[141] S. Nolfi and D. Floreano. Coevolving predator and prey robots: Do arms races arise in artificial evolution? *Artificial Life*, 4:311–335, 1998.

[142] S. Nolfi and D. Floreano. *Evolutionary Robotics. The Biology, Intelligence, and Technology of Self-Organizing Machines*. MIT Press, 2001.

[143] P. Nordin, W. Banzhaf, and M. Brameier. Evolution of a world model for a miniature robot using genetic programming. *Robotics and Autonomous Systems*, 25:105–116, 1998.

[144] M. J. North, N. T. Collier, and J. R. Vos. Experiences creating three implementations of the repast agent modeling toolkit. *ACM Transactions on Modeling and Computer Simulation*, 16, 2006.

[145] M. Okura, A. Matsumoto, H. Ikeda, and K. Murase. Artificial evolution of fpga that controls a miniature mobile robot khepera. In *SICE 2003 Annual Conference*, volume 3, pages 2858 –2863, 2003.

[146] Y. Ong, P. Nair, and A. Keane. Evolutionary optimization of computationally expensive problems via surrogate modeling. *AIAA Journal*, 41:687–696, 2003.

[147] I. Paenke. *Dynamics of Evolution and Learning*. Phd thesis, Universität Karlsruhe (TH), 2008.

[148] G. Parker and R. Georgescu. Using cyclic genetic algorithms to evolve multi-loop control programs. In *IEEE International Conference on Mechatronics and Automation*, pages 113 – 118, 2005.

[149] F. Pasemann, U. Steinmetz, M. Hülse, and B. Lara. Evolving brain structures for robot control. In *Connectionist Models of Neurons, Learning Processes, and Artificial Intelligence*, pages 13–15. Springer, 2001.

[150] T. Peters. Pep 20 – the zen of python. *Python Enhancement Proposals*, 20, 2004.

[151] A. D. Pietro, L. While, and L. Barone. Applying evolutionary algorithms to problems with noisy, time-consuming fitness functions. In *Congress on Evolutionary Computation*, volume 2, pages 1254 – 1261, 2004.

[152] M. Pigliucci. Is evolvability evolvable? *Nature Reviews Genetics*, 9:75–82, 2008.

[153] W. po Lee, J. Hallam, H. H. Lund, and E. U. K. Applying genetic programming to evolve behavior primitives and arbitrators for mobile robots. In *In Proceedings of the 4th International Conference on Evolutionary Computation*, pages 495–499. IEEE, 1997.

[154] A. Prügel-Bennett and A. Rogers. *Modelling genetic algorithm dynamics*, pages 59–85. Natural Computing Series. Springer, 2001.

[155] M. Quinn, L. Smith, G. Mayley, and P. Husbands. Evolving team behaviour for real robots. In *International Workshop on Biologically-Inspired Robotics: The Legacy of W. Grey Walter*, pages 14–16, 2002.

[156] S. Railsback, S. Lytinen, and V. Grimm. Stupidmodel and extensions: A template and teaching tool for agent-based modeling platforms, 2005. http://condor.depaul.edu/slytinen/abm/StupidModel, retrieved on November 11th 2014.

[157] S. F. Railsback, S. L. Lytinen, and S. K. Jackson. Agent-based simulation platforms: Review and development recommendations. *Simulation*, 82:609–623, 2006.

[158] T. Ray. An approach to the synthesis of life. *Artificial Life II, Santa Fe Institute Studies in the Sciences of Complexity*, 11:371–408, 1991.

[159] Real Time Engineers Ltd. FreeRTOS – the standard solution for small embedded systems, 2012. http://www.freertos.org, retrieved on November 11th 2014.

[160] I. Rechenberg. *Evolutionsstrategie : Optimierung technischer Systeme nach Prinzipien der biologischen Evolution*. Number 15 in Problemata. Frommann-Holzboog, 1973.

[161] I. Rechenberg. Case studies in evolutionary experimentation and computation. *Computer Methods in Applied Mechanics and Engineering*, 186:125 – 140, 2000.

[162] T. Reenskaug. Thing-model-view-editor – an example from a planningsystem, May 1979. http://heim.ifi.uio.no/~trygver/1979/mvc-1/1979-05-MVC.pdf, retrieved on November 11th 2014.

[163] T. Reenskaug. Models-views-controllers, Dec. 1979. http://heim.ifi.uio.no/~trygver/1979/mvc-2/1979-12-MVC.pdf, retrieved on November 11th 2014.

[164] J. Reisinger, K. Stanley, and R. Miikkulainen. Towards an empirical measure of evolvability. In *Proceedings of the 2005 workshops on Genetic and evolutionary computation*, pages 257–264. ACM, 2005.

[165] H. G. Rice. Classes of recursively enumerable sets and their decision problems. *Transactions of the American Mathematical Society*, 74:358–366, 1953.

[166] C. Ronnewinkel, C. O. Wilke, and T. Martinetz. Genetic algorithms in time-dependent environments. In *Theoretical Aspects of Evolutionary Computing*, pages 263–288. Springer, 1999.

[167] M. Rubenstein and R. Nagpal. Kilobot: A robotic module for demonstrating behaviors in a large scale (2^{10} units) collective. In *Proceedings of the IEEE international conference on robotics and automation*. Institute of Electrical and Electronics Engineers, 2010.

[168] S. J. Russell and P. Norvig. *Artificial Intelligence: A Modern Approach*. Pearson Education, 2003.

[169] B. Sareni and L. Krahenbuhl. Fitness sharing and niching methods revisited. *IEEE Transactions on Evolutionary Computation*, 2:97 –106, 1998.

[170] F. Schlachter, E. Meister, S. Kernbach, and P. Levi. Evolve-ability of the robot platform in the symbrion project. In *IEEE International Conference on Self-Adaptive and Self-Organizing Systems*, pages 144–149, 2008.

[171] A. C. Schultz, J. J. Grefenstette, and W. Adams. Roboshepherd: Learning a complex behavior. In *Proceedings of the Florida Artificial Intelligence Research Symposium*, pages 763–768, 1996.

[172] A. Serenko, B. Detlor, and Business. *Agent toolkits: A general overview of the market and an assessment of instructor satisfaction with utilizing toolkits in the classroom*. Michael G. DeGroote School of Business, McMaster University, 2002.

[173] P.-O. Siebers and U. Aickelin. Introduction to multi-agent simulation. *Computing Research Repository*, 2008.

[174] E. Sklar. NetLogo, a multi-agent simulation environment. *Artificial Life*, 13:303–311, 2007.

[175] E. Sober. The two faces of fitness. In *Thinking about Evolution: Historical, Philosophical, and Political Perspectives6*, 2001.

[176] D. A. Sofge, M. A. Potter, M. D. Bugajska, and A. C. Schultz. Challenges and Opportunities of Evolutionary Robotics. In *International Conference on Computational Intelligence, Robotics and Autonomous Systems*, 2003.

[177] W. M. Spears and D. F. Gordon. Evolution of strategies for resource protection problems. In *Advances in evolutionary computing: theory and applications*, pages 367–392. Springer, 2000.

[178] V. Sperati, V. Trianni, and S. Nolfi. Evolving coordinated group behaviours through maximisation of mean mutual information. *Swarm Intelligence*, 2:73–95, 2008.

[179] R. K. Standish. Going stupid with EcoLab. *Simulation*, 84:611–618, 2006.

[180] K. O. Stanley and R. Miikkulainen. Competitive coevolution through evolutionary complexification. *Journal of Artificial Intelligence Research*, 21:63–100, 2004.

[181] K. Sterelny. What is evolvability? *Handbook of the Philosophy of Science – Philosophy of Biology*, 3:177 – 192, 2006.

[182] T. Stützle, M. López-Ibáñez, P. Pellegrini, M. Maur, M. M. de Oca, M. Birattari, and M. Dorigo. Parameter adaptation in ant colony optimization. In *Autonomous Search*, pages 191–215. Springer, 2012.

[183] Sun-Microsystems and Oracle. Java website, 2014. http://www.java.com/en, retrieved on November 11th 2014.

[184] M. Szymanski. *Entwicklungsumgebung für Roboterschwärme*. Phd thesis, Karlsruhe Institute of Technology, 2011.

[185] M. Szymanski, T. Breitling, J. Seyfried, and H. Wörn. Distributed shortest-path finding by a micro-robot swarm. In *Ant Colony Optimization and Swarm Intelligence*, pages 404–411. Springer, 2006.

[186] M. Szymanski, J. Fischer, and H. Wörn. Investigating the effect of pruning on the diversity and fitness of robot controllers based on mdl2e during genetic programming. In *IEEE Congress on Evolutionary Computation*, pages 2780–2787. IEEE Press, 2009.

[187] M. Szymanski, L. Winkler, D. Laneri, F. Schlachter, A. van Rossum, T. Schmickl, and R. Thenius. SymbricatorRTOS: A flexible and dynamic framework for bio-inspired robot control systems and evolution. In *IEEE Congress on Evolutionary Computation*, pages 3314–3321. IEEE Press, 2009.

[188] A. Thompson. Evolving electronic robot controllers that exploit hardware resources. In *Advances in Artificial Life: Proceedings of the 3rd European Conference on Artificial Life*, pages 640–656. Springer, 1995.

[189] R. Tobias and C. Hofmann. Evaluation of free java-libraries for social-scientific agent based simulation. *Journal of Artificial Societies and Social Simulation*, 7, 2004.

[190] V. Trianni. *Evolutionary Swarm Robotics*, volume 108 of *Studies in Computational Intelligence*. Springer, 2008.

[191] V. Trianni and M. Dorigo. Self-organisation and communication in groups of simulated and physical robots. *Biological Cybernetics*, 95:213–231, 2006.

[192] V. Trianni and S. Nolfi. Engineering the evolution of self-organizing behaviors in swarm robotics: A case study. *Artificial Life*, 17:183–202, 2011.

[193] A. M. Turing. Computing machinery and intelligence. *Mind*, 59:433–460, 1950.

[194] J. Urzelai and D. Floreano. Incremental evolution with minimal resources. In *First International Khepera Workshop*, 1999.

[195] J. Urzelai, D. Floreano, M. Dorigo, and M. Colombetti. Incremental robot shaping. *Connection Science*, 10:341–360, 1998.

[196] L. G. Valiant. Evolvability. *Electronic Colloquium on Computational Complexity*, 6, 2006.

[197] G. P. Wagner and L. Altenberg. Complex adaptations and the evolution of evolvability. *Evolution*, 50:967–976, 1996.

[198] M. Waibel, L. Keller, and D. Floreano. Genetic team composition and level of selection in the evolution of cooperation. *IEEE Transactions on Evolutionary Computation*, 13:648–660, 2009.

[199] J. Walker, S. Garrett, and M. Wilson. Evolving controllers for real robots – a survey of the literature. *Adaptive Behavior*, 11:179–203, 2003.

[200] R. Watson, S. Ficici, and J. Pollack. Embodied evolution: Distributing an evolutionary algorithm in a population of robots. In *Robotics and Autonomous Systems*, pages 1–18, 2002.

[201] K. Weicker. *Evolutionäre Algorithmen*. B.G. Teubner, 2007.

[202] D. Weyns, H. V. D. Parunak, F. Michel, T. Holvoet, and J. Ferber. Environments for multiagent systems state-of-the-art and research challenges. In *Environments for Multi-Agent Systems*, pages 1–47, 2004.

[203] D. Weyns, M. Schumacher, A. Ricci, M. Viroli, and T. Holvoet. Environments in multiagent systems. *The Knowledge Engineering Review*, 20:127–141, 2005.

[204] D. Weyns, A. Omicini, and J. Odell. Environment as a first class abstraction in multiagent systems. *Autonomous Agents and Multi-Agent Systems*, 14:5–30, 2007.

[205] D. White, A. Arcuri, and J. Clark. Evolutionary improvement of programs. *IEEE Transactions on Evolutionary Computation*, 15:515–538, 2011.

[206] S. Wischmann and F. Pasemann. The emergence of communication by evolving dynamical systems. In *From animals to animats 9: Proceedings of the Ninth International Conference on Simulation of Adaptive Behaviour*, 2006.

[207] J. Zagal, J. Ruiz-del Solar, P. Guerrero, and R. Palma. Evolving visual object recognition for legged robots. In *RoboCup 2003: Robot Soccer World Cup VII*, volume 3020 of *Lecture Notes in Computer Science*, pages 181–191. Springer, 2004.

[208] J. Ziegler and W. Banzhaf. Evolving control metabolisms for a robot. *Artificial Life*, 7:171–190, 2001.

[209] J.-C. Zufferey, D. Floreano, M. v. Leeuwen, and T. Merenda. Evolving vision-based flying robots. In *Proceedings of the Second International Workshop on Biologically Motivated Computer Vision*, pages 592–600. Springer, 2002.

Index